T0282463

SpringerBriefs in Electrical and Computer Engineering

More information about this series at http://www.springer.com/series/10059

Iosif I. Androulidakis

VoIP and PBX Security and Forensics

A Practical Approach

Second Edition

 Springer

Iosif I. Androulidakis
Pedini Ioannina
Greece

ISSN 2191-8112   ISSN 2191-8120   (electronic)
SpringerBriefs in Electrical and Computer Engineering
ISBN 978-3-319-29720-0   ISBN 978-3-319-29721-7   (eBook)
DOI 10.1007/978-3-319-29721-7

Library of Congress Control Number: 2016934056

Printed on acid-free paper

This Springer imprint is published by Springer Nature
The registered company is Springer International Publishing AG Switzerland

To my parents

Preface

Apart from the public telephone network we all know, there is a parallel private network, consisting of private branch exchanges (PBXs). These are privately owned telephone exchanges that serve the communication needs of a private or public entity making connections among internal telephones and linking them to other users in the public telephone network.

Modern societies rely on telecommunication infrastructure more than ever. PBXs serve Hospitals, Ministries, Police, Army, Banks, Public bodies/authorities, Companies, Industries, and so on. This leads to the assumption that most—if not all—of the nations' vital infrastructures rely on PBXs as well. As such, it is not an exaggeration to state that PBXs are part of a nation's critical infrastructure. The purpose of this second edition of the book, therefore, is to raise user awareness in regard to security and privacy threats present in PBXs, helping both users and administrators safeguard their systems. Morover, this second edition has an extended coverage on VoIP systems.

It is focused on practical issues and easy-to-follow examples, skipping theoretical analysis of algorithms and standards. The book is more geared towards the telephony as a service and the devices themselves and not the underlying networks, so most of the contents are applicable to PSTN and VoIP alike. The contents are balanced, including both technical and nontechnical chapters. Amateur as well as experienced administrators will benefit from the overview of threats and the valuable practical advice. They will also get to know various issues affecting the security of their PBX while they will also learn the fraudsters' modus operandi. More advanced administrators will appreciate the technical discussions and will possibly try experimenting with the forensics and PBX control techniques presented in the respective chapters.

Chapter 1 gives an introduction to PBXs and the scene, statistics, and involved actors. Confidentiality, integrity, and availability threats are discussed in Chap. 2 providing the background for the highly technical discussion of Chap. 3. Having examined the threats and the technical background, Chap. 4 deals with security. Forensics involving PBXs are covered in Chap. 5. Concluding the book, Chap. 6 synopsizes the previous chapters.

Closing, I would like to thank my family for all the support and love, my profes-
sors in Greece and Slovenia for their mentoring during my studies, and the security
researchers all over the world with whom I have met and collaborated. There are too
many to be listed here but they know who they are! Last but not least, I would like
to thank my Editor and all the members of the Springer team with whom I have
collaborated. I hope you will like this book as much as I enjoyed writing it.

Ioannina, Greece   Iosif I. Androulidakis, Ph.D., Ph.D.
October 2015

Contents

Chapter 1
Introduction

1.1 About PBXs

Apart from the public telephone network we all know, there is a parallel private network, consisting of private branch exchanges (PBXs).[1] These are privately owned telephone exchanges that serve the communication needs of a private or public entity making connections among internal telephones and linking them to other users in the public telephone network. They exist in the form of IP PBXs (using the IP protocol via VoIP technologies) and conventional time division multiplexing (TDM) PBXs (using phone lines). Their software can be offered proprietary or via open source. While communication with other entities takes place using trunk lines to the public telephone network (or the Internet), internal telephone traffic fully depends on PBXs.

Initially, the primary advantage of PBXs was cost savings on internal phone calls: handling the circuit switching locally reduced charges for local phone service. As PBXs gained popularity, they started offering services that were not available in the operator network, such as hunt groups, call forwarding, and extension dialing. Since PBXs incorporate telephones, fax machines, modems, and even computers, the general term "extension" is used to refer to any end point on the branch. A typical PBX is depicted in Fig. 1.1. Opening the rack doors, Fig. 1.2 depicts the front side and Fig. 1.3 the back side.

At this point, we have to make an important clarification. The interested reader might have noticed that the PBX term is an old term, mainly referring to the TDM technology. The one-way trend of the past few years is to move to IP and Telephony convergence using Voice over IP technologies and products. Clearly we are in the VoIP era; however, the part of the book that analyses PBXs manages to stay current for a number of reasons explained below.

---

[1] Other uses of the PBX acronym include frightening terms such as polymer-bonded eXplosive, plastic-bonded eXplosive and pre-B-cell leukemia homeoboX!

© Springer International Publishing Switzerland 2016
I.I. Androulidakis, *VoIP and PBX Security and Forensics*, SpringerBriefs
in Electrical and Computer Engineering, DOI 10.1007/978-3-319-29721-7_1

Fig. 1.1 A typical PBX

The technology change towards VoIP has mostly focused on the lower levels (physical to transport) where now data packets flow in IP networks rather than voice samples in circuit switched telephone lines and trunks. The core functionality and the interworking have remained the same, and as such most of the dangers and manipulation techniques that will be examined in this book still apply for VoIP. Needless to say that new attacks focusing on VoIP explicitly have of course emerged [1]. We will thus include discussion on VoIP security.

VoIP is a descendant of computer and networking science rather than telecommunication science. While computer and network security has been extensively researched and awareness levels have rose, the same has not happened with PBXs. This is why this contribution tries to bridge this gap. Traditional PBXs have been exceptionally rigid and enjoy life spans of more than 20 years. As such, infrastructure that was commissioned in early 2000, before the major VoIP "invasion," will stay in operation for another 10 years from now, reinforcing the need for specific literature. In addition to that, lots of installations were based on hybrid solutions before migrating to full VoIP. Manufacturers in any case kept large segments of code and functionality from older tested and proven PBX platforms. This way, older problems migrated to the new platforms.

In any case, fraudsters utilize almost the same modus operandi in their attacks, and the end results are the same dangers in confidentiality, integrity, and availability. At the end of the day, given the global crisis, old installations will hang on for quite long since they have proven their value and rigidness.

Fig. 1.2 The front side of the PBX

1.2   PBXs as Critical Infrastructure

According to EU, Critical infrastructure is defined as:

> The physical and information technology facilities, networks, services and assets that, if disrupted or destroyed, would have a serious impact on the health, safety, security or economic well-being of citizens or the effective functioning of governments in EU countries.[2]

---

[2]Council of the European Union. 2008. "COUNCIL DIRECTIVE 2008/114/EC of 8 December 2008 on the identification and designation of European critical infrastructures and the assessment of the need to improve their protection." European Union. (http://europa.eu/legislation_summaries/justice_freedom_security/fight_against_terrorism/l33260_en.htm).

Fig. 1.3 The back side of the PBX with cables going to the distribution frame

Modern societies rely on telecommunication infrastructure more than ever. Former US president Bill Clinton reconfirmed that telecommunication networks have entered the critical infrastructure (CI) domain long ago [2]. Economy, health, industry, security, private and public sector have extended telecommunication networks based on PBXs in order to serve their communication needs. There are millions of lines installed in every country and they essentially complement the public network. PBXs serve Hospitals, Ministries, Police, Army, Banks, Public bodies/authorities, Companies, Industries, and so on. This leads to the assumption that most—if not all—of the nations' vital infrastructures rely on PBXs as well.

Even if the core public network is operating normally, unintentional or targeted damages and attacks in PBXs can cause significant instability and problems as well as significant financial losses [3–9]. Furthermore, interception of calls is a very sensitive issue that affects all of us. Clearly, PBX attacks can be used in warlike situations. Effectively bringing down PBXs puts a nation's critical infrastructures at risk. The lack of understanding and acting upon the interdependencies of critical infrastructures [10] can lead to unexpected cascading effects. Taking all the previous into account, it is not an exaggeration to state that PBXs are part of a nation's critical infrastructure. As such, the "scene" is quite active.

1.3 The Scene

PBX system suppliers include well-known brands that share the largest part of the market, plus smaller ones. It is hard to deliver definite numbers of PBX installations in the market, as the penetration calculations are complicated. Thousands or even millions of installations are present in most countries, serving many times more users.

As with every other technology, the rate of promotion and marketing of new products, operations, and services in the market is overwhelming. In addition to that, there is extensive complexity and significant systems interaction. It is therefore logical that a thorough testing of all possible scenarios and parameters that can lead to vulnerabilities is a process that cannot be applied with absolute success. This technical difficulty in combination with the stringent time margins and deadlines to launch the products and services makes things worse. The result is of course gaps and inefficiencies in regard to security in all aspects.

At the same time, telecommunication fraud is as old as the first manual telephone exchanges back in early twentieth century. CFCA (Communications Fraud Control Association), in a 2012 survey [11] announced that the annual global fraud losses are in the range of $72–$80 billion for 2008 up 34% from the results of 2005. Twenty-seven percent of companies reported losses greater than 5% of their revenue. Among the top three fraud loss categories, 20% of losses (~$15 billion) is attributed to compromised PBX/Voicemail Systems. Luckily, thanks to the increase in collaboration and coordination among carriers in identifying and stopping fraudulent activity, 2015 fraud losses have been cut to $38.1 billion down 18% from 2013. As a percent of global telecom revenues, fraud losses are approximately 1.69%—a 0.40% decrease from 2013. At the same time the top method for committing fraud is still PBX Hacking (classical or VoIP PBX) [12].

An older survey (2003) of CFCA [13] finds PBX/Voicemail fraud and Calling Card fraud prevailing but most interestingly gives a possible link to terrorism. As the survey states: "With respect to the causes of the growth of telecom fraud, some telecom providers did report that global fraud losses had partly risen due to an increase in worldwide terrorism. Terrorist organizations embrace telecom fraud to generate funds by illegally gaining access to a network and then reselling the service." A more recent case proves that there is indeed such a connection [14]. Apart from the economic impact, the link of PBX abuse to terrorism is an aspect that should well be taken into account.

PBX manufacturers along with users, administrators, and telecom providers–carriers and their employees are all responsible for the security. Checking the proper operation and ensuring the safety of PBX as well as the protection against unauthorized use and access are usually left to the owner. This has of course tremendous effects since due to economic and technical difficulties, in essence it is impossible to guarantee that the proper measures are taken. Manufacturers themselves in the respective user manuals state that it is impossible to guarantee 100% security since the owner has the final word in setup and administration of the switch and as such every unauthorized use claims are charged to the owner. Taken verbatim from a

PBX user's manual [15]: "No telecommunications system can be entirely free of risk from unauthorized use …. Because the customer has ultimate control over the configuration and use of services and products purchased, the customer properly bears responsibility for fraudulent uses of those products and services."

1.4   The Players

The main actors involved in cyber-attacks (such as PBX attacks are) can be split into attackers and targets. Within these two groups we distinguish several categories. In regard to attackers, their motivation varies. There are those who do it for fun (or holding a grudge, such as a laid off employee), those who do it for financial reasons (to save money or make profit), those who do it for political reasons, and those who do it to raise their profile.

The scale of importance grows from low to extremely high. At the lower end of the impact scale are skilled individuals, usually teenagers trying to break in just for the challenge or fun. Traditionally, people who studied, experimented with, and explored telecommunication systems were called "phreakers," the term being a portmanteau of the words phone and freak [16]. Nowadays, the most common threat to a network is the malicious hacker, or groups of them, that share their findings and coordinate their efforts. They are trying to earn personal benefits by employing their skills to deploy various illegal activities.

Organized crime has its own customer base that demands cheap international calls and will break into PBXs to serve this base, either with low profile and long duration operations or, more commonly, with intense exploitation in a few days that leaves the targeted company with a huge bill. They will also use the PBX as an anonymizing communication link.

Spies have also lots to gain from intruding into PBXs. They can intercept phone calls and logs, financial and technical data providing valuable information, especially in cases of industrial espionage. This espionage can be ordered by a foreign nation or a competitor, and it is very possible that an internal link in the company of the organization, usually an employee, is used.

Dishonest employees and administrators can resort to various ways of manipulating the PBX, to make free calls, to elevate their calling privileges or alleviate the barring of calls, and so on. Insiders are probably the most difficult enemy to deal with. They can prove a valuable ally for a malicious hacker, providing him with passwords and information about the infrastructure. They could also simply give him permission to enter a company and poke around the equipment. Finally, insiders could act by themselves exploiting company's assets, planting "bugs" and interception devices, and so on. For example, an employee, contractor, or even a cleaner could forward a seldom-used extension to an overseas number and make international calls by calling a local rate number in the office.

At the other end of the spectrum are terrorists who according to recent surveys seem to manipulate telephone exchanges in order to raise funds for their purposes as stated earlier. Finally, let us not forget the physical disasters. A flood or a fire can destroy the PBX and consequently deprive the company from telephone service for a long period.

On a cyber conflict perspective, attackers can be state actors, state sponsored actors, or non-state actors. Non-state actors can be split into cyber criminals, revolution groups and terrorists, and script kiddies. State actors and state sponsored actors most likely operate in a hierarchical structure. Non-state actors are likely to be less structured and can operate in the Cell or Forum format—working either on their own or using a Forum to collaborate. Script kiddies rely heavily on the Forum format. While cyber criminals, revolution groups, and terrorists can operate in the Forum format, these groups can also work in powerful cells which make these groups more unpredictable [17].

The skills essential to any successful team of attackers are imagination and creativity, while the Internet allows attacks to take place anonymously and asymmetrically [2]. Technical skills and knowledge on how PBXs operate are essential. They need however to be complemented with nontechnical ones such as social engineering [5]. Given the many undocumented features and commands found in proprietary systems, insider's help from manufacturers could be an extra asset.

Finally, between the attackers and the PBX targets, the legitimate users of the PBX (including administrators) are found. As previous research has shown, users are the weak link in security. Users themselves are critically affected by security and privacy threats, and play a key role in protecting themselves and others. Since users are not adequately informed in regard to the threats they face, they do not actively follow most of the security best practices. They are also not familiar with all of the functions the PBX offers and their security independencies. Therefore, the problem is not only technical, but it also extends to users' education and awareness.

1.5 Conclusion

With such an active "scene" and so many "players," it is a matter of time before vulnerabilities are discovered (and possibly exploited), from researchers in the best case or from criminals in the worst case. PBXs are not an exception to this rule.

Much has been said and done regarding data communication security but PBXs have always been almost forgotten. Being unprotected they are an easy target to be attacked and as such PBX fraud has been sustained for a long period of time due to ignorance and naivety. With the advent of new telecommunication technologies which are based around open communications via the Internet (VoIP), the situation gets even more complicated.

PBXs are part of a nation's critical infrastructure and attacks in their confidentiality, integrity, and availability can have a serious impact at any given time. Unfortunately, telephony security has never gotten near the levels of security IT systems enjoy. The stats on PBX abuse and telecom fraud are overwhelming, proving that the "players" involved in the scene are more active than ever.

Academia and industry should focus on their security awareness campaigns and efforts in order to combat the attackers. In the following chapter we will focus on specific confidentiality, integrity, and availability threats helping raise users' and administrators' awareness.

References

1. Walsh TJ, Kuhn DR (2005) Challenges in securing voice over IP. IEEE Secur Priv 3(3):44–49
2. Geers K (2011) Strategic cyber security. CCD COE, Tallinn, Estonia
3. West D (2000) De-mystifying telecom fraud. Telecom Business, July 2000
4. Blake V (2000) PABX security, information security technical report, vol 5, no 2, pp 34–42
5. Mitnick KD, Simon WL (2002) The art of deception: controlling the human element of security. Wiley, Indianapolis
6. Androulidakis I (2010) Detecting cybercrime in modern telecommunication systems. In: European Police College (CEPOL), Seminar 64/2010, Cyber Crime & High Tech, Athens, 18–21 May 2010
7. Androulidakis I (2011) Cybercrime in fixed telephony systems. In: European Police College (CEPOL), Seminar 62/2011, High Tech & Cyber Crime, Brdo near Kranj, Slovenia, 20 Oct 2011
8. Archer K, White GB et al (2001) Voice and data security. Sams Publishing, Indianapolis
9. Pollard C (2005) Telecom fraud: the cost of doing nothing just went up, White paper. Insight consulting, Feb 2005
10. Luiijf E, Klaver M (2011) Insufficient situational awareness about critical infrastructures by emergency management. TNO Defence, Security and Safety
11. CFCA, Communications Fraud Control Association (2009) Worldwide Telecom Fraud Survey
12. CFCA (2015) Global Fraud Loss Survey. http://www.cfca.org/pdf/survey/2015_CFCA_Global_Fraud_Loss_Survey_Press_Release.pdf
13. CFCA, Communications Fraud Control Association (2003) Worldwide Telecom Fraud Survey
14. Ars Technica (2011) How Filipino phreakers turned PBX systems into cash machines for terrorists. http://arstechnica.com/tech-policy/news/2011/11/how-filipino-phreakers-turned-pbx--systems-into-cash-machines-for-terrorists.ars
15. Avaya Inc (2002) Avaya products security handbook, issue 8, Chap. 2
16. Wikipedia, Phreaking. http://en.wikipedia.org/wiki/Phreaking
17. Ottis R (2011) Theoretical offensive cyber militia models. In: Proceedings of the 6th international conference on information warfare and security, Washington, DC

Chapter 2
Confidentiality, Integrity, and Availability Threats in PBXs

2.1 Introduction

Applied practice and specific incidents have shown that PBXs suffer from a series of problems that negatively influence the confidentiality, integrity, and availability of communications. Truth is that the academic community has not shown great interest in the PBXs' security, possibly because of their traditionally closed and proprietary nature. Luckily, with VoIP, things changed and there is a revived research interest.

Despite the increased involvement of the academic community in the security discussions, many of the traditional problems (especially the ones that are targeting devices, implementations, and services) have not been solved in VoIP, but rather new ones have emerged targeting even the protocols themselves.

Apart from technical issues, the user remains one of the weakest links in the security ecosystem. This is why this contribution mainly aims at raising user awareness. In this introductory chapter, we will briefly describe and group as many as possible of the threats in confidentiality, integrity, and availability that PBXs are facing. This way, it will be the basis for the technical discussions that will follow in the next chapters. It will also help better understand how to apply the security measures presented in the respective chapter later on.

2.2 Confidentiality

Eavesdropping is of course what comes first on mind when talking about telephony confidentiality threats. Voice communications can be intercepted in real time or recorded to be retrieved and analyzed later. Calls can be multiplexed to internal or external listening posts. They can also be rerouted to other destinations (possibly utilizing a man in the middle attack too). Apart from phone voice calls, fax calls and (given the convergence of telephony and Internet) data traffic can be intercepted and extracted.

© Springer International Publishing Switzerland 2016
I.I. Androulidakis, *VoIP and PBX Security and Forensics*, SpringerBriefs
in Electrical and Computer Engineering, DOI 10.1007/978-3-319-29721-7_2

Fig. 2.1 Serving area interface

In general, eavesdropping communications makes organizations dealing with confidential information liable. In addition to this, industries can be confronted with espionage problems. PBX hacking can lead to the interception of strategic or tactical information of the army, police, or ministries as was also proven in warlike situations [1].

Even in digital PBXs where switching is performed in a digital manner, low end phones can still be analog ones, where the audio is carried in an analog way from the PBX board to the telephone set. A simple tap connected in the cable leading to the phone (everywhere in its possible route, from the main distribution frame in the provider's premises, to the serving area interface, to the local distribution frame up to inside the telephone itself) can easily extract the audio or provide access to the line to make free calls. This is very well known as "clip-on" fraud. Figure 2.1 shows a serving area interface and Fig. 2.2 a close-up of a distribution frame. We can notice sensitive information regarding telephone numbers and routing and patching information left in small pieces of paper by the technicians. More importantly, a tool

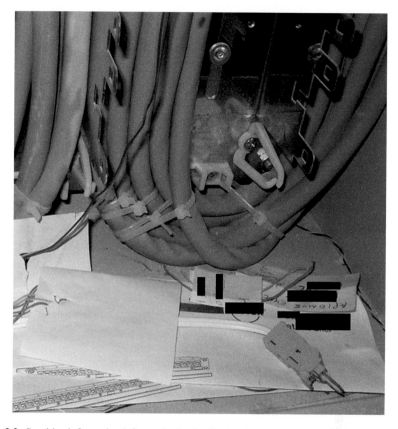

Fig. 2.2 Sensitive information left over in the distribution frame plus the tool to intercept calls

to intercept calls is seen, which when inserted in the respective connections automatically intercepts the audio without disconnecting the line at all.

In end to end digital systems, as well as in VoIP systems audio is traveling as digitized packets. If there is no encryption, these data packets can be translated back to audio. Specifically for PBXs there exist special hardware modules that perform the task for most brands, translating the proprietary digital signals to audible analog audio (Fig. 2.3).

Telephone sets, on the other hand, can be modified to transfer voice while on hook, with either hardware modification or software commands. Indeed, it is completely impossible to force a $10 analog phone to do something without a hardware modification, while a $200 digital phone with all these bell and whistles can (easily) be reprogrammed or commanded from the PBX to switch its microphone on, and relay the audio heard in the surrounding environment to the attacker. It transforms this way to a very efficient remotely controlled bug.

A classical way of interception is the use of special interception devices, the well-known "bugs." The topic of Technical Surveillance Counter Measures for

Fig. 2.3 Board translating proprietary digital audio to analog (after http://www.trx.com.pl/)

defense against electronic eavesdropping and bugging is a whole topic by itself, so we will not address it in this book. We will, however, promptly mention some information on interception devices here.

Interception devices may be placed on the telephones we use or on the cabling and the network and provider's equipment. The traditional method of clip-on, where a pair of wires (and possibly a phone) is connected in parallel with the original one, is still the most effective one! The presence of voltage on the wire itself allows the continuous and long-term operation of a transmitter bug without the need for replacing batteries such as Fig. 2.4 depicts. The latter can even be placed inside the telephone device itself. The picture of a movie spy who unscrews the handset's microphone on an old analog telephone device is very typical even though the technology used today is different. Another basic element that is worth mentioning is that of a device installed in a telephone device that can also intercept ambient conversations taking place inside the office and not solely during calls. This device can be "waken-up" and start relaying audio from the microphone of the phone, without the phone ever ringing.

Wiretaps that either record or transmit the voice are more common but it is also possible to conduct a passive interception without any kind of intervention. Signal interception from wires is possible, through induction. In Fig. 2.5 a special inductive coil is shown which intercepts telephone signals without breaking the circuit, just by picking up the electromagnetic fluctuations. The basic principle here is that according to Maxwell's equations, a varying electric field produces a magnetic field and vice versa. Physical audio in electronic devices is transformed to a varying electrical current that produces a magnetic field. So, the induction coil picks the magnetic field, transforms it back to current, and drives an amplifier to allowing overhearing the initial audio.

Fig. 2.4 Wireless transmitter for telephone conversations

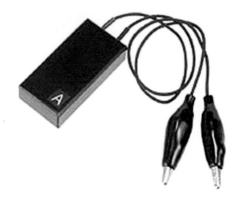

Fig. 2.5 Inductive coil for intercepting telephone signals

A more elaborate technique is that of "the man in the middle." In order to inter-cept communications a hacker can intervene in the middle pretending (from a tech-nical point of view) to be the other party and relaying the information to the intended party. Revealing sensitive information during a phone call is thus never a good idea unless some sort of cryptographic means is used.

Interceptions can take place in an active and/or passive way. In the active method, there is some kind of physical or technical interaction with the target phone (or network) and as such it is easier to be spotted. Active ways of interception include, e.g., connect-ing extra cables in the main distribution frame (clip-on) or reprogramming the PBX.

Passive interceptions, on the other hand, are very frequent in wireless systems, by just "listening" to the airwaves and decrypting the encrypted radio wave com-munications that flow in the air. They are focused on cryptanalysis, and can target protocols and algorithms in various parts of the network. In a typical wireless com-munication, the radio waves are transmitted freely in the air and they cannot be easily confined. Therefore, a potential intruder can intercept and process these sig-nals without even coming close to the target. As a matter of fact, older, analog wire-less calls could passively be intercepted using cheap gear since there was no kind of encryption at all. Digital wireless systems changed that. Up to recently, it has been traditionally difficult to intercept air waves for DECT cordless phones that are also used in PBXs. However, some open source projects during the few past years have brought this functionality to the interested user with great success [2].

Leaving aside malicious actions regarding interception of messages, there is also the lawful interception where operators are required to help law enforcement authorities. Each provider–carrier is obligated to offer mechanisms to monitor communications for the needs of authorities, following the necessary warrant. Lawful interception functionality abuse can prove to be one of the most effective spying options. Although it is a very sensitive system that should be adequately protected, successful attacks have been seen [3]. Employees with the proper authorization credentials or malicious hackers that manage to break in the system (with or without internal help) can intercept at will the calls of the targets they chose.

Administrators can also activate monitoring and debugging features that allow real-time interception. There is also software offered by 3rd party companies for the same goal. Most PBXs offer tools to silently listen to other calls, in the context of a supervisor checking the agent's performance and behavior. It is also part of the secretary–manager association, where the secretary can intrude and inform the manager about an urgent call. Both systems can be abused.

Usually, when an intrusion takes place an informatory beep in regular time intervals (e.g., every 15 s) informs the parties that the call is monitored. This is a very important point. Unfortunately, very few people realize that this tone is actually a warning about the presence of another party in the conversation. Even worse, administrators can disable this tone, or set it to a very short duration or to a nonaudible tone (very high or very low frequency).

It is not always necessary to eavesdrop on the actual communication-voice call. The interception of call logs could reveal many useful connections. With call correlation and traffic analysis techniques, interesting connections can be revealed (such as the contacts of the user) and industrial espionage information gathered even without knowing the actual content of the calls, but only the identities of the parties. As an example, if the logs reveal a flow of calls between a company and the patent office, a safe assumption is that a new product is on its way to be patented. Respectively, key suppliers can be found and so on.

Among the harvested information threatening the confidentiality would be forwarded numbers in the forward tables as well as mobile phones and personal numbers present in the logs and stored in the memories of the phones and the PBX. Many users are also saving personal codes such as PINs in the memory of the phone, so these could also leak. Finally, imagine a fraudster, having the ability to intercept credit card numbers as the unsuspected client presses the keys in his phone to perform a telephone-banking transaction.

2.3 Integrity

The second major part in the taxonomy of threats deals with integrity attacks. Such attacks on a PBX vary from reprogramming it, installing backdoors for future access, altering data (e.g., erasing incriminating information), and modification of features (enable services that are banned) to economic fraud. Deleting or changing the respective call and operation logs would cover the entry and the trails of the attackers. At the same time, it is possible to alter the communication flow, connecting lines to different destinations (possibly utilizing a man in the middle attack). Harming the integrity of data in specific files and memory locations can lead to denial of service and PBX shutdown.

As experienced administrators can attest, it is easier to find the cause of a malfunction when it manifests as a complete lack of service than it is to find out what is wrong when the system misbehaves or when the error appears sporadically. Thus, an attacker can create more damage by forcing the PBX to function incorrectly rather than launching a full denial of service attack. Changing settings is sometimes more effective than completely shutting service altogether. For example, randomly forwarding, swapping numbers, or changing caller IDs among users can cause havoc and it simultaneously takes more time to realize what is wrong.

Identification of the caller to the calling party is one of the handiest features of telephony. Of course, the ability to withhold the caller ID from being presented to the called party has been associated with an array of unpleasant acts, such as harassments, pranks, spam calls, calls from unwanted persons, and so on. Furthermore, having access to a PBX, the identity of the sender can be changed in order to make a malicious call appear legitimate or for spam purposes. This call spoofing takes place when a caller masquerades as another one. It is used to avoid paying for the service, to hide the fact that the call is an unsolicited or fraudulent one, and to avoid detection in case of attacking further nodes. Call spoofing is very important in cases of Vishing. This is a practice where fraudsters gain access to private personal and financial information by convincing their victims they are calling from their bank. Counting in the trust people have to telephone service, fraudsters are using the proper caller ID of a bank. Indeed, telephony numbers have traditionally been connected to physical locations and companies, so users tend to trust a telephone number, thinking of it as a valid form of ID.

Staying with the integrity analysis we will also examine two nontechnical effects of integrity attacks in PBXs, psyops and fraud. Psyops is short for psychological operations, which can be defined as "Planned operations to convey selected information and indicators to foreign audiences to influence their emotions, motives, objective reasoning, and ultimately the behavior of foreign governments, organizations, groups, and individuals" [4]. Examples of such incidents would be to play propaganda messages to all internal users (they would listen to them as soon as lifting the handset). It could also connect callers and incoming lines to recorded messages. Messages could be played by PAs (public address system–loudspeakers) connected to the PBX. Furthermore, constantly ringing phones or phones ringing in the middle of the night could cause severe discomfort, annoyance, or even fear.

Apart from Psyops, as telephony security is usually lacking compared to IT security, the opportunities for crime are numerous. Unauthorized access of telephone service with the relevant economic losses due to bills is a major concern. Losses due to computer incidents are usually given only as an estimate and it is indeed a very complex procedure yielding wrong results many times. Economic losses due to a telephony incident on the other hand are immediately obvious. As was seen in the

```
<CHFLME:?

Change follow me relation

CHFLME: <FM-TYPE>,<ORIG-DNR>sr[,<DEST-NUMBER>];

If DEST-NUMBER is omitted the follow me relation of ORIG-DNR will be erased.

<CHFLME:0,101,102;
EXECUTED
<CHFLME:0,101;
EXECUTED
<CHFLME:0,201,00███████;
EXECUTED
```

Fig. 2.6 Setting up a call forward to an external number in a command line interface

introduction, the costs attributed to telecom fraud, and specifically PBXs are not easily appreciated reaching $72 to $80 billion [5]. Imagine a telephony fraud taking place unnoticed for a substantial period. Apart from the apparent cost of the bill, lost revenues and additional expenses can skyrocket the total loss to astronomical amounts. Money laundering through PBXs is also effective, while consequences after a multimillion fraud in calls starting from a company's PBX would lead to financial and business disaster. Even terrorist organizations are thought to be embracing telecom fraud to generate funds [6, 7].

Fraudsters abuse telecom services (using stolen codes and access and charging third party users), allowing use of the infrastructure by non-authorized persons, making free calls or selling these calls via a call-selling operation, making a profit. Call-selling operations to high cost international destinations can prove to be very profitable when promoted to certain groups of people who have increased communication needs, such as students, refuges, foreigners, military personnel, and imprisoned ones. In case the company owns a free calling, 800 number, then it is even more alluring for the attacker since even the initial call (from a payphone usually) is free. Moreover, a simple forwarding of a seldom-used extension to an overseas number can be easily commanded, as shown in Fig. 2.6.

Fraudsters can actually steal expensive boards of the PBX (e.g., the CPU board) and parts of the infrastructure that can be later sold in the black market. Luckily, in some brands, expensive boards such as the CPU board are protected by means of cryptographic hashes. A side effect of this would be the resulting denial of service as will be discussed later on. In a lighter tone, an attacker can control a whole PBX in order to win in a contest where players are calling in a Radio or TV station. This was achieved by Kevin Poulsen; hacking the KIIS-FM 102 station's PBX he managed to be the 102nd caller as required by the competition, wining this way a Porsche 944 [8].

Compromised PBXs calling premium rate services (that can explicitly be set just for the purposes of receiving calls from these infiltrated PBXs) or high cost destinations such as overseas or satellite networks can lead to extensive bills. Consider 100 compromised PBXs in a country, each calling such services or destinations 1,000 times per day. Assuming a cost of $10 per call yields a total of $1 million per day in

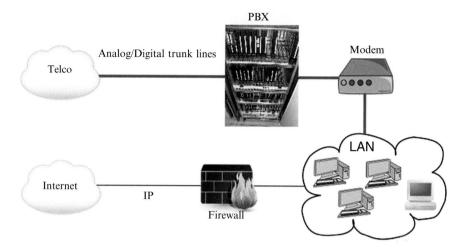

Fig. 2.7 Firewall bypass due to a non-authorized modem

bills! The attacker could even be aiming at calls that are terminated in another country. This way, local operators would owe the foreign ones a substantial amount due to call termination rates leading to significant economic losses for the country. A side effect would be the international circuits blocking as described in the denial of service section.

Although oxymoron, it seems that low end and cheaper PBXs are less vulnerable than the larger PBXs with richer functionality. The configuration of a small PBX is practically impossible to be attacked (in order to intercept or change the integrity of data) from outside since such PBXs do not offer any remote access to the administration at all! As a matter of fact, their programming usually takes place by dialing special codes issuing the respective commands, using a phone already connected in the PBX. Even with the absence of remote administration, however, they are still subject to denial of service attacks, internal fraud, and/or hardware modifications.

Closing the integrity section, it must be noted again that the PBX can be the weak link to target the IT platform that is interconnected with it (either for administrative only or for full service reasons such as the case with VoIP). When a PBX is linked to an organization's IT network, malicious hackers can find an easier point of entry (in case the PBX maintenance port is not adequately secured) into the critical assets, customer databases, and business applications of the target.

There is also another great risk in regard to PBX and IT interconnection: Some users are connecting unauthorized modems in their phone lines in order to bypass enforced Internet access policies. They do that by connecting their PCs to the telephone network via the PBX (instead of using VPNs and corporate secured and tested solutions). As shown in Fig. 2.7, the upper part of the connection bypasses the firewall, providing access to the corporate network via a modem. Employees are even using remote access programs to work from home. An attacker could find these modems using war dialing (as will be explained in the next chapter), enter their PCs, and jump to the IT network, bypassing firewall security and further penetrating the network as an internal user. There are many cases where a perfectly well-designed computer network is brought down due to errors and omissions in the telephone network. It is rather oxymoron to invest into computer security but to forget to invest into telephone security. This is why total security can only be achieved with combined efforts and supplies between IT and telecom world.

2.4   Availability

Integrity implies there is availability in the first place. There is no point discussing about integrity when a system does not work at all! Given telephony's omnipresence, its availability has been considered a given fact. Truth is, however, that it is relatively easy to block PBX communications. A lack of effective telephone communication can cause annoyance for its users [9]. More importantly, the service of hospitals can be damaged by phones which are not working [10]. It is frightening to imagine not being able to call a hospital in an emergency. Furthermore, national economy could suffer great losses if a targeted attack was to render useless industry's telecommunication lines. In any case, it is apparent that in the modern demanding business environment a company or organization cannot survive without telephone service.

The fact that telephony network outages can have serious consequences became clear by the 9/11 terrorist attacks on the World Trade Center. The telephony network for first responders was lost rapidly, and due to failing backups the problems with the network continued. This led to more difficulty with arranging rescue and recovery [11]. At the same time, the inability to reach beloved ones can cause stress for many people, which opens a possibility for deliberately bringing down networks to perform psyops of the same nature during warlike or in harassment situations.

A new form of denial of service is connected to credit card fraud. Fraudsters are blocking the phone lines of victims as to inhibit the bank from informing them about abnormal behavior concerning their credit card. This is arguably more a problem for household lines rather than PBXs. Both, however, might be using wireless phones (practically DECT sets, since analog wireless phones have long ago ceased to exist). As with every other wireless technology, DECT handsets can be victims of jamming attacks that mask and overcome in power the legitimate signals rendering phones unusable.

Proceeding, an attacker can affect the availability of the system for both management and communication itself, affecting administrators and the internal and external users. He could cut off administrator's access by changing passwords, disabling remote access, and shutting connectivity to IP and serial port communications. Incoming and outgoing lines could be administratively set out of service, effectively isolating the PBX from the outside world. More aggressively, deleting the database which contains the setup of the PBX or files from the operating system could completely halt the switch. Interestingly, at least one manufacturer has specific login accounts with the sole purpose of halting the PBX or deleting and reinstalling the settings database, effectively wiping the existing setup. With the same logic, it is possible to overwrite the firmware of the boards of the PBX using its firmware update tools. Finally, a simple shutdown command could be ordered. Placed in a UNIX-like cron file, this keeps the PBX shutting down at predetermined time intervals.

Another effective denial of service technique can target the software and hardware protection of PBXs. Modern PBXs employ protection against unauthorized copying and black market selling utilizing some form of hardware key, usually with a field programmable gate array (FPGA) integrated circuit. The FPGA holds a specific hash that is coupled to the hardware present in the PBX. If somebody tampers with this protection mechanism, the PBX will enter a limited functionality mode and eventually will shut down, protecting the vendor's profits and the PBX from illegal interventions and black market operations. Specifically for PBXs exported to third countries, the hardware key can only be shipped from the manufacturing country since local dealers do not have access to it, further extending the time needed to restore its functionality.

It is not always needed to have physical access in a PBX in order to harm its availability. A remote denial of service in the communication abilities can be achieved by another PBX (or an array of PBXs) attacking the target with hundreds of calls per minute. This attack could be used against the better-protected PBX where the attacker could not obtain access. It effectively blocks the incoming and outgoing lines and the legitimate users cannot utilize them anymore since they are constantly busy. As an example, a typical E1 PRI European ISDN line offers 30 voice channels (23 with T1 lines used in USA). Assuming short calls of 10 s, including the setup time, a rate of 180 calls per minute can be achieved against the target. A medium size PBX with three E1 lines can launch 540 calls per minute against a given target. This would easily overwhelm most small PBXs. In a full ISDN environment with call setup times of less than 1 s, the whole duration of the call could be as little as 2 s, yielding a 2,700 calls per minute rate from the same three E1 setup. But, even if the target has enough capacity in lines, it could be well possible that the shear rate of calls causes a bottleneck in another part of the system, e.g., the automated attendant (press 1 for sales, 2 for accounting, etc.) or the Interactive Voice response platform (say sales to get to sales, accounting for accounts, etc.). This blocking of lines scenario closely resembles the denial of service and the distributed denial of service attacks of the computer networks, where instead of packets, calls are now flooding it. In any case, a classical computer denial of service could target the administrative computer platform. Along the same line, a possibly more annoying way of attacking the availability of a given phone is to place consecutive calls to it, effectively making it ring nonstop. The user is then forced to unplug it.

At this point, another interesting effect of the repeated calls attack could be the overloading of international circuits. Indeed, the international circuits' capacity is usually limited. With an array of PBXs calling random destinations abroad, the links could be saturated and the country cut off from the international network. In order for the attack to fully succeed, it should be launched against PBXs that are being served from all possible international connectivity providers–carriers. Since communication circuits are two-way, the attack would have the same effect using compromised PBXs in the target country only, performing outgoing calls, or compromised PBX in other countries performing incoming calls towards the target country.

Fig. 2.8 Close-up of boards

Returning to hardware, a non-confirmed and as far as the author knows not researched, yet somehow plausible, attack on PBXs is the following. By intervening in the firmware of the PBX, it could be possible to force the PBX to perform functions that could lead to physical failure of its electronic components. Especially for analog lines that employ relays, an attack could force the relays into a constant rapid on–off loop that would ultimately burn them or it could force the ringing signal to permanently be applied in a phone, possibly burning its ringer circuit. Looped reboots or file access could also wear the hard disc of the PBX. The consequences of such attacks are far less severe compared to damages caused to an industrial system by similar means (such as Stuxnet caused). None the less, these attacks can still be implemented to bring down a PBX.

Another factor that often slips our attention is the probability of theft of (parts of the) PBX. Modern telephone exchanges use expensive and easily removed and carried equipment (e.g., exchange cards—boards, Fig. 2.8), so a couple of minutes would be enough for an incident to take place causing apart from the economic damage also an outage. At the same time, this board can host a memory chip or a hard disc storing valuable information (in an unencrypted way most usually). As a matter of fact, the theft can be only temporary, lasting for a few minutes. It is only that long it takes for a malicious person to install software or tamper with the electronic circuits.

Finally, let us not forget that interruptions in the service are not always caused by attackers. More than often, technical glitches, bugs, or environmental disasters cause extended scale and duration service interruption incidents. This is why protection against environmental elements and disasters is essential.

2.5   Other Threats

Apart from the confidentiality, integrity, and availability issues, PBXs face other threats too that combine all the previous. A common use of a compromised telephone network is to use it as a screen for covering-up criminals' illegal activities such as ring operations, drug selling, money laundering, etc. A covert call usually originates from payphones because they can offer anonymity and they are easy to find and accessible from almost everywhere. Further anonymity is granted while the call is routed through many PBXs to make it extremely difficult to trace. This "looping" is a very effective way to mislead authorities from tracing them. The technique was on the decline with the advent of convenient prepaid mobile phones but is expected to catch up again, since new legislation requires prepaid mobile phones to be register with valid IDs before being used. This anonymity can be exploited to attack other targets, making the compromised PBX an intermediate point and its owner possibly held liable for the attack.

At the same time, the compromised PBXs (and especially voice mail systems) can be a repository of illegal information to be exchanged by the fraudsters. Encrypted messages could be stored by criminals and retrieved by their peers (e.g., for drugs dealing) while multiparty calls can take place, organizing actions.

A side effect (or a deliberate one) is the adverse publicity and the damage to the reputation of a company that has been hit by such an incident. Indeed, customers could be denied telephone service, or a disgruntled ex-employee can forward company's incoming lines to sex lines. Even worse, clients could be connected to an adversary's line. A fitting example here comes from the early days of telephony. Almon Strowger invented an automatic telephone exchange because he was convinced that the wife of a competitor being an operator at a manual telephone exchange was sending calls to her husband instead of him whenever someone asked for the service they both offered.

Speaking of adverse publicity, an ever growing percentage of emails in the Internet are spam. The same can eventually happen with voice calls. Spam could be defined as unsolicited commercial messages from unknown originators. It is well possible that a compromised PBX can take place in such an act, by relaying audio messages to randomly chosen telephone numbers. The user gets a call and at the other end of the line, the prerecorded message starts playing passing political propaganda or trying to convince him to visit a website to hopefully make a transaction buying a good or service.

So far we have examined technical threats. However, it is not always necessary to be technically savvy to abuse a telephone network. A very common technique for accessing it is the use of social engineering; people who pretend to be someone else use their persuasion to extract valuable information for the network itself or information that can be helpful for infiltrating it. There are two good examples here, one is the use of social engineering by a person that impersonates a trusted one (e.g., an employee) via the phone and extracts information from a secretary, a username and a password to login to the network, and the other is a person who gives false information and impersonates a network technician in order to extract information about the whereabouts of the PBX. Gaining this way the guard's approval to access the PBX he has full access to the network. Further examples of social engineering can be found in [12].

2.6 Specifically for VoIP

As already stated, most threats already discussed affect VoIP the same way they affect classical TDM PBXs. Voice over IP Security Alliance (VoIPSA) [13] has nonetheless provided a specific security threat taxonomy. It includes social threats, eavesdropping, interception, and modification threats, service abuse, intentional interruption of service (including denial of service), and other interruptions of service.

Social threats exploit vulnerabilities of VoIP in regard to caller ID and the lack of a "physical" bond of a given number to an actual address. Eavesdropping counts on unencrypted traffic that can be "sniffed" all the way along the network, much easier than actually clipping-on cables in an underground cable. Indeed, there are dozens of tools for decoding video and audio streams. Interception and rerouting of calls are possible by targeting specific vulnerabilities in the protocols and the signaling during call setup. Denial of service can be achieved not only by direct attacks to VoIP protocols and implementations but also by attacking the supporting computer network infrastructure (traffic flooding, attacks in the DNS servers, etc.).

In any case, VoIP is implemented using a range of components and protocols that all have potential vulnerabilities, bugs, and exploits as presented in [14].

- VoIP Elements

    - Call control elements (call agents)

        Appliance or server-based call control—Internet protocol private branch exchange (IPPBX)
        Soft switches
        Session border controllers (SBCs)
        Proxies

    - Gateways and gatekeepers

        Dial peers

    - Multi-conference units (MCUs) and specialized conference bridges
    - Hardware endpoints

        Phones
        Video codecs
        Other devices and specialized endpoints

    - Soft clients and software endpoints

        IP phones
        Unified messaging (UM) integrated chat and voice clients

Desktop video clients
IP-based smartphone clients

– Contact center components

Automated call distribution (ACD) and interactive voice response (IVR) systems
Call center integrations and outbound dialers
Call recording systems
Call center workflow solutions

– Voicemail systems

• Protocols

– H.248 (Megaco)
– Media gateway control protocol (MGCP)
– Session initiation protocol (SIP)
– H.323
– Skinny call control protocol (SCCP) and other proprietary protocols
– Session description protocol (SDP), real-time protocol (RTP), real-time control protocol (RTCP), and real-time streaming protocol (RTSP)
– Secure real-time transport protocol (SRTP)
– Inter-Asterisk eXchange protocols (IAX and IAX2)
– T.38 and T.125
– Integrated services digital network (ISDN)
– Signaling system number seven (SS7) and SIGTRAN
– Short message service (SMS)

The point to be made is that despite the trends and the changes in technology from TDM PBX to VoIP PBX, threats are still the same, the methods are more or less the same, and only technical details in specific attacks change. We will shed some more light on the technical details in the next chapter.

2.7 Conclusion

It is more than clear that PBXs face an increased array of threats that affect their availability, confidentiality, and integrity. Apart from voice communication, enhanced capabilities and services including phone banking transactions raise even stronger security issues.

Although modern technology security area is a constant battleground, telephony security levels are not high enough. This presents a vast opportunity for telecom carriers and service providers too. They can play a proactive and strategic role in protecting their subscribers, both through education and through the security software they should deploy across their networks.

Manufacturers on the other hand should proceed to better-designed interfaces and systems generally, richer in security features. Special software could help users mitigate some of the security risks offering embedded encryption options as well as automated backup features and options to detect fraud.

To combat the threats, before rushing to technological solutions the first step could be the awareness and better education of users. Following this introduction, we will continue with a more technical chapter before presenting security solutions.

References

1. Akhvlediani M (2009) The fatal flaw: the media and the Russian invasion of Georgia. Small Wars Insurgencies 20(2):363–390
2. deDECTed. https://dedected.org/trac
3. Prevelakis V (2007) The Athens affair. IEEE Spectrum, July 2007
4. The Free Dictionary (2012) Psychological operations. http://www.thefreedictionary.com/psychological+operations
5. CFCA, Communications Fraud Control Association (2009) Worldwide Telecom Fraud Survey
6. CFCA, Communications Fraud Control Association (2003) Worldwide Telecom Fraud Survey
7. Ars Tsechnica (2011) How Filipino phreakers turned PBX systems into cash machines for terrorists. http://arstechnica.com/tech-policy/news/2011/11/how-filipino-phreakers-turned-pbx--systems-into-cash-machines-for-terrorists.ars
8. Brisbane (2006) http://www.bris2600.com/hall_of_fame/poulsen.php
9. BBC News (2011) Exeter hospital phones 'cannot understand Devon accent'. http://www.bbc.co.uk/news/uk-england-devon-14649238
10. St. Cloud Times Minnesota (2011) BRIEF: fire causes problem for hospital phones. St. Cloud, Minnesota. http://www.newsorganizer.com/article/brief-fire-causes-problem-for--ac0a083204b6cc060f0c2d0edede85fa/
11. Luiijf E, Klaver M (2011) Insufficient situational awareness about critical infrastructures by emergency management. TNO Defence, Security and Safety
12. Mitnick KD, Simon WL (2002) The art of deception: controlling the human element of security. Wiley, Indianapolis
13. VOIPSA (2005) VoIP security and privacy threat taxonomy public release 1.0, 24 October 2005. http://www.voipsa.org/Activities/VOIPSA_Threat_Taxonomy_0.1.pdf
14. Rhodes-Ousley M (2013) Information security: the complete reference, 2nd edn. McGraw-Hill Education, New York

Chapter 3
PBX Technical Details

3.1 The PBX Basic Structure

In this chapter we will focus on technical details, describing the parts of a PBX and how each one can affect its security. Figure 3.1 will be our basic reference for the discussion. Before getting to the PBX itself, we will examine the connection to the public network and to other PBXs. We will also see the cabling and distribution frames and then move into the PBX itself, describing boards, sets, services, etc.

3.2 Connection to the Outside World

Starting with the connection to the outside world, it can be achieved with a number of technologies and protocols such as ISDN E1/T1, Analog, CAS 2bit, IP, GSM/3G (with FCTs), etc. The medium used can be copper, optical fibers, and wireless technologies including microwave, WiFi, infrared, etc. The "frontier" of the PBX is the demarcation point, that is, the point that interconnects the customer's network with the external network. Figure 3.2 shows the demarcation point of a large installation.

In cases of PBX networks, other nodes can be interconnected, usually with specific universal interconnection protocols such as DPNSS, QSIG, or proprietary ones (e.g., ABC). The medium options are of course the same as with the outside world connections. As an example, Fig. 3.3 presents an optical patch panel with optical links to the rest of the interconnected PBXs.

© Springer International Publishing Switzerland 2016
I.I. Androulidakis, *VoIP and PBX Security and Forensics*, SpringerBriefs
in Electrical and Computer Engineering, DOI 10.1007/978-3-319-29721-7_3

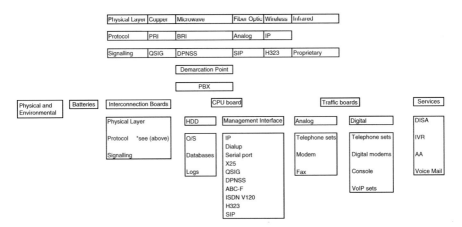

Fig. 3.1 PBX structure

3.3   Distribution Frames-Cabling

In order for the PBX to be connected to the public network, to another PBX, or even to its internal users, a network of cables and distribution frames is needed. Distribution frames along with the cabling are the "circulatory system" of the whole infrastructure. The main distribution frame is the part where all internal and external lines are connected to the PBX, the "heart" of this network of cables. It is placed close to the PBX, with one part of it dedicated to the lines coming from the PBX and the other part of it dedicated to the lines leaving towards the users and the external network. Figures 3.4, 3.5, and 3.6 show different size distribution frames. Cables going to the internal users can be terminated in intermediate distribution frames, e.g., in every floor of the building where they are placed in racks such as one pictured in Fig. 3.7.

The route from the PBX to the set can be many hundreds of meters long, with cables passing through all these distribution frames and ducts. This makes it possible to insert an interception device anywhere in the length of this route, or cross connect phones in parallel, transferring voice from the victim to the listening post. Indeed, as Fig. 3.8 depicts, it is very difficult to locate a "suspicious" cable among the dozens of cables fitted in such a small area. Earlier in Chap. 2, we saw in Figs. 2.1 and 2.2 examples of points where a "bug" can be installed.

3.4   Physical Parameters

Continuing with physical PBX parameters, we have the physical location where the PBX is actually located in. Attacks can be facilitated by lax physical security measures. There are many cases where the PBX is placed in the basement or in the corridors of an organization, behind unlocked doors, or even worse, in plain view. There are even cases

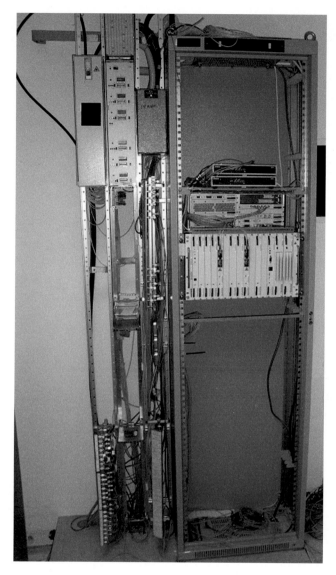

Fig. 3.2 Large size office demarcation point

where informatory signs and labels point to the PBX. Phones in public areas such as the elevator phone or the operator's console are immediate targets.

The physical access, for the daring fraudsters, involves actually visiting the PBX or its cabling and performing their malicious actions on-site. As a matter of fact, it need not be a stealth operation. The fraudster can use social engineer tricks, posing as a technician, as maintenance personnel, as a colleague (in a big company with hundreds of employees), and even be shown the way to the PBX from helpful (but not security

Fig. 3.3 Optical patch panel

Fig. 3.4 Small office main distribution frame and CPE

Fig. 3.5 Medium size office main distribution frame and CPE

educated) employees. Achieving physical access the attacker can perform hardware modifications or connect to the management port. It is also possible to use a console station already on place. In a more advanced level, physical access in equipment could provide material for reverse engineering (e.g., of boards or telephone sets).

In absolute numbers, the administrator must have a clear image of the number of sets, number of trunk lines, and central office lines as well as the number of cabins, shelves, boards, and other elements that compromise the PBX. Electrical parameters including consumption and the state of UPS and batteries are also essential for the operation of the PBX. Diesel generators can be in place in critical installations such as in hospitals. Finally, environmental parameters include temperature and humidity with fire and flood protection measures in place. In any case, the whole area should be kept clean and tidy.

We won't extend any more in order to focus on more PBX specific issues, starting with the boards.

3.5 PBX Boards and Hardware

Getting to PBXs themselves we notice that most of them are designed using a modular architecture. There is a chassis offering backbone connectivity as was seen in Fig. 1.2, and various boards are inserted in slots, each board performing a specific action. Figure 3.9 depicts the back side of this chassis.

Fig. 3.6 Large company main distribution frame

There is the CPU board, boards to serve analog phones, digital phones, trunk lines, and so on. It must be noted that in PBX terms, there is some sort of "spatial–geographical" discrimination. Each rack, shelf, board, and connection of the board has a number. Figure 3.10 shows a listing of boards, as returned from the respective command issued in the management console of the PBX. As an example, there is a board named UA32 in this specific brand, placed in the slot 1–22.

The CPU board is usually the most expensive and complex one. In any case almost every single board holds memory chips and microprocessors, while boards serving telephones (usually an even number) show a distinctive pattern that allows a specific phone to be targeted for hardware intervention. Figure 3.11 shows the specific UA32 board that was just mentioned. It is a board serving 32 digital extensions. The user can clearly distinguish 32 physical blocks in the upper left side, one for each phone, and 16 blocks in the upper right and lower left side, each one serving two extensions.

Fig. 3.7 Rack with patch panels

3.6 PBX Sets

The main "users" of the PBX are its telephone extensions. A multitude of sets can be connected in a PBXs, given the respective boards exist. A typical example of a (partial) listing of sets along their calling numbers is seen in Fig. 3.12. The board of Fig. 3.11 that was mentioned in previous section (3.5) is inserted in slot 1–22 and it has 32 extensions–numbers. The extensions it serves are appointed "spatial–geographical" IDs 1-22-0 to 1-22-31, each ID mapped to a given extension-number. The connection from the boards to the sets can be wired or wireless (e.g., DECT), with one pair of two pairs of cables, an optical link or even a distance extension device (repeater).

Fig. 3.8 Close-up of Fig. 3.6

Fig. 3.9 Back side of the chassis

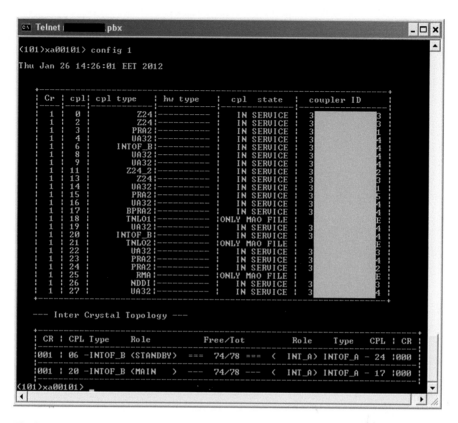

Fig. 3.10 List of boards (hardware configuration) installed in a PBX and their status

The "spatial–geographical" representation of PBX elements that we already mentioned is quite universal in PBXs. Figure 3.13 shows an example from another brand. Querying the details of extension with internal number 4491 the administrator can see that it is hosted in slot 1-0-41-3.

There are classical analog sets and many different digital ones with an LCD or not ranging in functionality and cost (the more buttons the more expensive!). These sets have different physical, electric, and logical characteristics. Leaving aside an attack to the PBX itself, analog sets need a physical intervention in order to be bugged (should be dismantled to have electronic bugs installed, or have their speed dialing memories read). Digital ones, full of microprocessors, EPROMS, and ICs, are susceptive to both hardware and software attacks. In any case, the goal of the attacker would be to intercept information, not only during a call but also discussions held in the surrounding area of the phone, by making the phone transfer voice while on hook while at the same time it appears as being idle, innocently waiting for calls.

Another sensitive point in regard to telephone sets consists of the internal phones placed in publicly accessed areas (e.g., in the lobby or in the elevator). As a matter of fact there is also a whole category in relevant articles in underground electronic magazines regarding what is called "elevator phreaking," Such phones are easy to

Fig. 3.11 A PBX board serving 32 digital sets

access and as an internal part of the network can easily be misused to expose vulnerabilities. Internal phones are furthermore a convenient access point for all those who want to harm the network and its infrastructure. An intruder can easily install a "bug" or use them just to place a free call. He can also set a forward to an external number. As such, publicly accessed phones have to be both protected and confined in places that access is subject to some kind of monitoring and control. In case they are really needed, all necessary steps must be taken in order to secure them and make sure that they cannot cause problems.

A special case of an internal phone is the operator's console (Fig. 3.14). If not properly administered, it may have the ability to change PBXs setup features and operational data. It could for example unblock barred destinations or leverage call abilities on certain phones. It could also initiate call forwarding in different sets. A very powerful and dangerous feature is the busy override or intrude feature. Using it, the operator can intrude in a call taking place and intercept it. Usually there are warning tones in the communicating parties that inform them about the intrusion. These tones can however be disabled, set to a very short duration or to a very high or low, inaudible frequency. The users on the other hand rarely are informed about the meaning of these tones, so they might not get alerted. There is also the obvious threat of making calls if the operator has not logged-out or locked the console before leaving. The status of the console is checked in a given brand using the respective commands as shown in Fig. 3.15.

Other special phones not from a technical point of view but rather based on the importance of the user are the phones of executives. Targeting the phone of a CxO could reveal his personal mobile phone number or other personal information such

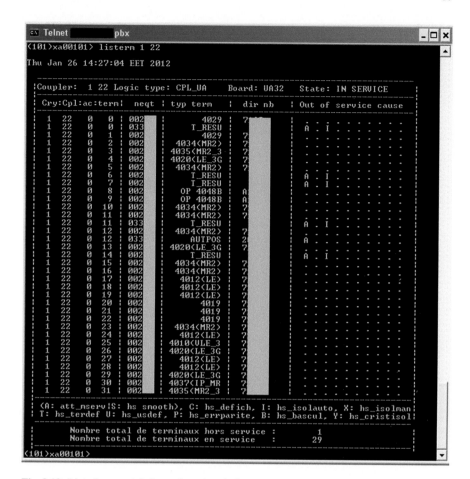

```
CV Telnet ▓▓▓▓▓ pbx                                                  _ □ ✕
<101>xa00101> listerm 1 22                                               ▲
Thu Jan 26 14:27:04 EET 2012

---------------------------------------------------------------------------
:Coupler:  1 22 Logic type: CPL_UA    Board: UA32   State: IN SERVICE   :
:--------------------------------------------------------------------------:
: Cry:Cpl:ac:term:  neqt : typ term   : dir nb : Out of service cause   :
:  1  22  0   0 : 002 :       4029 : ? :  . . . . . . . . . . :
:  1  22  0   0 : 033 :     T_RESU : ? : A . I . . . . . . . . :
:  1  22  0   1 : 002 :       4029 : ? :  . . . . . . . . . . :
:  1  22  0   2 : 002 : 4034<MR2> : ? :  . . . . . . . . . . :
:  1  22  0   3 : 002 : 4035<MR2_3 : ? :  . . . . . . . . . . :
:  1  22  0   4 : 002 : 4020<LE_3G : ? :  . . . . . . . . . . :
:  1  22  0   5 : 002 : 4034<MR2> : ? :  . . . . . . . . . . :
:  1  22  0   6 : 002 :     T_RESU : : A . I . . . . . . . . :
:  1  22  0   7 : 002 :     T_RESU : : A . I . . . . . . . . :
:  1  22  0   8 : 002 :   OP 4048B : A :  . . . . . . . . . . :
:  1  22  0   9 : 002 :   OP 4048B : A :  . . . . . . . . . . :
:  1  22  0  10 : 002 : 4034<MR2> : ? :  . . . . . . . . . . :
:  1  22  0  11 : 002 : 4034<MR2> : ? :  . . . . . . . . . . :
:  1  22  0  11 : 033 :     T_RESU : : A . I . . . . . . . . :
:  1  22  0  12 : 002 : 4034<MR2> : ? :  . . . . . . . . . . :
:  1  22  0  12 : 033 :     AUTPOS : 2 : A . . . . . . . . . . :
:  1  22  0  13 : 002 : 4020<LE_3G : ? :  . . . . . . . . . . :
:  1  22  0  14 : 002 :     T_RESU : : A . I . . . . . . . . :
:  1  22  0  15 : 002 : 4034<MR2> : ? :  . . . . . . . . . . :
:  1  22  0  16 : 002 : 4034<MR2> : ? :  . . . . . . . . . . :
:  1  22  0  17 : 002 : 4012<LE> : ? :  . . . . . . . . . . :
:  1  22  0  18 : 002 : 4012<LE> : ? :  . . . . . . . . . . :
:  1  22  0  19 : 002 : 4012<LE> : ? :  . . . . . . . . . . :
:  1  22  0  20 : 002 :       4019 : ? :  . . . . . . . . . . :
:  1  22  0  21 : 002 :       4019 : ? :  . . . . . . . . . . :
:  1  22  0  22 : 002 :       4019 : ? :  . . . . . . . . . . :
:  1  22  0  23 : 002 : 4034<MR2> : ? :  . . . . . . . . . . :
:  1  22  0  24 : 002 : 4012<LE> : ? :  . . . . . . . . . . :
:  1  22  0  25 : 002 : 4010<VLE_3 : ? :  . . . . . . . . . . :
:  1  22  0  26 : 002 : 4020<LE_3G : ? :  . . . . . . . . . . :
:  1  22  0  27 : 002 : 4012<LE> : ? :  . . . . . . . . . . :
:  1  22  0  28 : 002 : 4012<LE> : ? :  . . . . . . . . . . :
:  1  22  0  29 : 002 : 4020<LE_3G : ? :  . . . . . . . . . . :
:  1  22  0  30 : 002 : 4037<IP_MR : ? :  . . . . . . . . . . :
:  1  22  0  31 : 002 : 4035<MR2_3 : ? :  . . . . . . . . . . :
:--------------------------------------------------------------------------:
: <A: att_mserv:S: hs smooth), C: hs_defich, I: hs_isolauto, X: hs_isolman:
: T: hs_terdef U: hs_usdef, P: hs_errparite, B: hs_bascul, Y: hs_cristisol:
:--------------------------------------------------------------------------:
:     Nombre total de terminaux hors service   :      1              :
:     Nombre total de terminaux en service     :     29              :
:--------------------------------------------------------------------------:
<101>xa00101>                                                            ▼
```

Fig. 3.12 List of sets and their numbers (*masked*) connected in a voice board

```
EXDDP : DIR=4491;

EXTENSION DIRECTORY DATA

DIR      CUST      EQU            CAT   TYPE    ICAT      AUX

4491      8        001 −0 −41 −03  5    EL6     0002

END
```

Fig. 3.13 Details of a single set

Fig. 3.14 A typical PBX operator's console

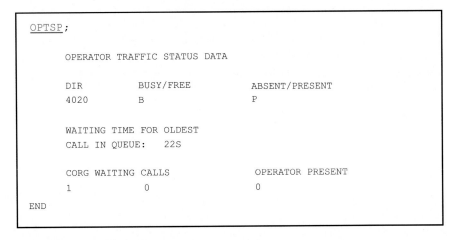

```
OPTSP;

        OPERATOR TRAFFIC STATUS DATA

        DIR             BUSY/FREE              ABSENT/PRESENT
        4020            B                      P

        WAITING TIME FOR OLDEST
        CALL IN QUEUE:     22S

        CORG WAITING CALLS             OPERATOR PRESENT
        1           0                      0
END
```

Fig. 3.15 Console status printout

as family call numbers. Given the fact that the more important a person in a company the more advanced a phone she has, the attacker could easily find out what are the numbers of these "important" persons judging by the set's type. CxO's would possibly have expensive phone sets with many keys and digital displays, while the rest of the employees could have simple analog phones or low end digital phones.

It must be noted at this point that modern systems save the numbers of the speed dialing entries in the PBXs' memory and databases rather than the phone itself. Older analog PBXs were relying on the phone's memory to store numbers. With that respect, older systems were more secure since the attacker would have to physically extract the data from the memory of the phone (possibly by stealing or replacing the target's phone). Now he can do the same by just reading the respective entries in the PBX's operating system, in a powerful and centralized manner.

Finally, there are VoIP sets too; those tend to have more "brains" than simple PBX telephones, and as such are more susceptible to malicious actions as we will see in the VoIP section.

3.7   The CPU and the Management Port

Serving all these sets (and not only), the "heart" of the PBX is the board hosting the CPU. Apart from the CPU, the same board hosts various memory circuits (ROMS, EEPROMS, Flash memories that contain voice guides), HDDs, a floppy disk, and serial or USB ports.

The CPU can be connected in a variety of ways to the outside world, in order for the PBX to be managed. There are direct or via modem serial port connections, proprietary protocols (usually linking to a digital set or the console), TCP/IP connectivity over Ethernet, V120 via ISDN, or even X25 in some older models.

The administration protocols for the management and interconnection of PBXs include PSTN and ISDN dialup over respectively analog or digital lines, generic networking protocols such as IP, X25, Frame-relay, and telephony-specific signaling protocols such as QSIG, DPNSS, SS7, SIP, and H323. There are also proprietary ones from different PBX vendors (e.g., ABC protocol). By monitoring these protocols an adversary can increase HIS target list finding further maintenance modem numbers and IP addresses from interconnected systems. In addition to that, potential targets can be found in modem logs, in routing and host tables and in extra subsystems and functionality present such as Voicemail. Figure 3.16 shows a screenshot from a PBX's hosts file, listing the close-by PBXs and CPUs.

As mentioned, a dialup line is usually used to connect the telephone exchange's CPU to the maintainer's modem in order to remotely administer the switch. This is one of the most dangerous features that can be misused causing not only telephone problems but also providing a way to enter the computer network. Figure 3.17 shows two modems one on top of the other. The upper one is a digital, ISDN modem while the bottom one is a classical analog modem. These boxes worth just $50 can lead to damages of millions of dollars since they effectively interconnect the PBX's management–maintenance port to the public telephone network, making it accessible from all over the world. All the attacker has to do is find their dial-in number and use the correct (usually the default) password. Having access to the switch, the attacker can reprogram it, turn on functions and services that can be exploited, and shut down other functionality such as call logging.

```
<106>xa00106> cat /etc/hosts
10.1.255.255      broadcast
127.0.0.1         loopback
#; Site
10.1.1.1          xa00101
10.1.1.2          xb00101
10.1.1.3          m00101
10.1.1.4          s00101
10.1.85.1         xa00101_C1
10.1.85.2         xb00101_C1
10.1.85.3         m00101_C1
10.1.85.4         s00101_C1
#; Site
10.1.8.1          xa00102
10.1.8.2          xb00102
10.1.8.3          m00102
10.1.8.4          s00102
10.1.92.1         xa00102_C1
10.1.92.2         xb00102_C1
10.1.92.3         m00102_C1
10.1.92.4         s00102_C1
#; Site
10.1.15.1         xa00103
10.1.15.2         xb00103
10.1.15.3         m00103
10.1.15.4         s00103
10.1.99.1         xa00103_C1
10.1.99.2         xb00103_C1
10.1.99.3         m00103_C1
10.1.99.4         s00103_C1
```

Fig. 3.16 Hosts file listing interconnected PBXs

Fig. 3.17 Typical modems

But how do hackers find the correct number to dial in order to connect to the maintenance port? They use a technique called "war dialing" which consists of calling every single number a company owns in order to discover modems and electronic services to abuse. The term comes from the 1983 classic film "War Games" that actually portrayed the technique. According to it, the attacker dials as many as possible numbers in a given range, trying to find modem carriers or other tones that

```
MS-DOS Prompt                                              _ □ ✕

🆃 9 x 15 ▾   ☐ ▤ ▤ ▣ ▤ ▤ A

        ToneLoc v0.99 (Beta-8) by Minor Threat & Mucho Maas (Mar 07 1994)

ToneLoc is a dual purpose wardialer.  It dials phone numbers using a mask that
you give it.  It can look for either dialtones or modem carriers.  It is useful
for finding PBX's, Loops, LD carriers, and other modems.  It works well with
the USRobotics series of modems, and most hayes-compatible modems.

USAGE:
ToneLoc  [DataFile]  /M:[Mask] /R:[Range] /X:[ExMask] /D:[ExRange] /C:[Config]
                     /#:[Number] /S:[StartTime] /E:[EndTime] /H:[Hours] /T /K

     [DataFile]  - File to store data in, may also be a mask       Required
     [Mask]      - To use for phone numbers    Format: 555-XXXX     Optional
     [Range]     - Range of numbers to dial    Format: 5000-6999    Optional
     [ExMask]    - Mask to exclude from scan   Format: 1XXX         Optional
     [ExRange]   - Range to exclude from scan  Format: 2500-2699    Optional
     [Config]    - Configuration file to use                       Optional
     [Number]    - Number of dials to make     Format: 250         Optional
     [StartTime] - Time to begin scanning      Format: 9:30p       Optional
     [EndTime]   - Time to end scanning        Format: 6:45a       Optional
     [Hours]     - Max # of hours to scan      Format: 5:30        Optional
                   Overrides [EndTime]
     /T = Tones, /K = Carriers (Override config file, '-' inverts)  Optional

C:\md5\toneloc>
```

Fig. 3.18 DOS era war dialing program

denote the presence of a computer/PBX. This is possible due to the direct inwards dialing (DID) or direct dial-in (DDI) service offered by all telecom providers, allowing an external user to reach a specific extension without the need to call the operator that would manually connect the call (and possibly deny connection to non authorized requests, at least if properly educated to do so).

As an example, consider a company listed in the yellow pages with the number 555-0000. Depending on the size of the company (whether having more than 100 internal phones or not), the attacker would dial all the numbers from 555-0001 to 555-0999, or from 555-0001 to 555-0099. To automate the tedious process of dialing, many programs were created, ranging from simple MS-DOS command line executables (Fig. 3.18) to full Graphical User Interface suites (Fig. 3.19) and from underground tools to legitimate commercial products. It can also be accomplished with manual dialing (finger dialing in the respective slang).

According to a recent research [1], war dialing leading to the maintenance modem of a PBX can be particularly effective (reaching even 70% success) by dialing only given extensions and not the whole range. In the previous example of 555-0000, the attack can yield immediate results by dialing number ending in 99, i.e., 555-0099, 555-0199, 555-0299 and so on. An older survey [2] regarding information security controls, "testing and review procedures including a 'war dial' of inbound phone lines to identify active modems" ranked last in a list of 80 controls. In other words, the identification and tracking of modem connections were incomplete, of low quality, and not rationalized posing a significant risk that shouldn't be neglected.

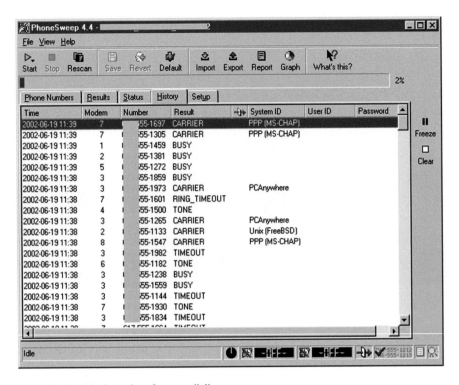

Fig. 3.19 Graphical user interface war dialing program

Fig. 3.20 Typical login
screen

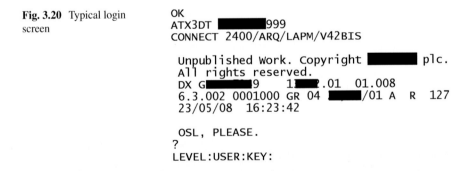

In any case, when connected, either directly via the serial port or via a modem or
IP connection, PBXs respond with login prompts such as the one in Fig. 3.20. An
interesting twist here is that some brands require a "wake up" sequence in order to
reply. This could be a sequence of characters (i.e., "ssssssssss"), or a special charac-
ter such as "ctrl-G".

This makes it easier to identify the given brand since most PBXs have distinctive
login screens. The attacker will then try default passwords on the target. It is very well
known that PBX maintenance platforms and ports, even more so than computers, tend

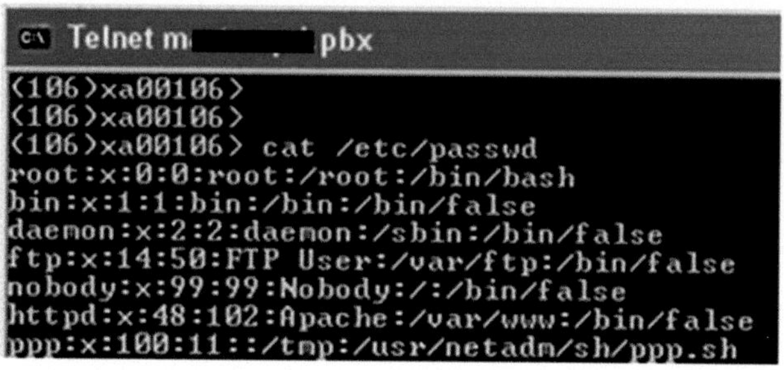

Fig. 3.21   Passwd file of a PBX

to have default passwords left. There are even long lists containing default ones [3]. Furthermore, there are cases of hardwired passwords in the code of PBXs which cannot be changed.

Furthermore, PBXs with UNIX-like operating systems maintain a passwd file, as seen in Fig. 3.21. This provides login names for which the respective passwords can possibly be brute forced. Brute forcing is an exhaustive method of systematically checking all possible passwords–keys until the correct one is found. As such it can be used in systems where no vulnerabilities are found. Luckily, this method would be of little success since most PBXs limit the password trials and after three or four wrong attempts disconnect the attacker.

It cannot be emphasized strongly enough that the dangers associated with remote access to the maintenance and management port enforce the need for competitive and trained local administration teams. Having a team dedicated to the management of the PBX minimizes the need for external connections and along with that the associated risks.

3.8   Software, Administration, and Management Suite and Station

The administration/management station can be a PC or server running whatever O/S. Smaller PBXs require the software to be installed in the external administration server, while larger PBXs typically include the software in their own operating system. The software running can be a closed proprietary operating system or based on generic operating systems specifically modified for the PBX. We will focus only on the management suite since the rest is a topic covered in computer security literature.

The management suite allows the provisioning of the PBX, controlling the operation of it, setting up, activating and modifying features, performing maintenance tasks, and so on. There are command line suites (Fig. 3.22), menu driven (Fig. 3.23), and full Graphical User Interface ones.

```
NANSI: NUMTYP=EX, NUMSE=1100;
EXECUTED




NANSI: NUMTYP=EX, NUMSE=1100;
EXTEI: DIR=1100, CAT=1, EQU=1-1-60-10, TYPE=EL6;
                                                EL6
                                                EL8
```

Fig. 3.22 Command line interface with auto-complete help

Fig. 3.23 Menu driven
PBX programming

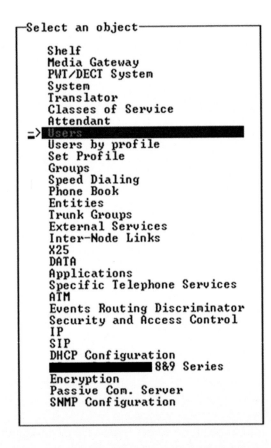

```
┌Select an object────────────────────────┐
│                                         │
│     Shelf                               │
│     Media Gateway                       │
│     PWT/DECT System                     │
│     System                              │
│     Translator                          │
│     Classes of Service                  │
│     Attendant                           │
│   =>█Users█████████████████████         │
│     Users by profile                    │
│     Set Profile                         │
│     Groups                              │
│     Speed Dialing                       │
│     Phone Book                          │
│     Entities                            │
│     Trunk Groups                        │
│     External Services                   │
│     Inter-Node Links                    │
│     X25                                 │
│     DATA                                │
│     Applications                        │
│     Specific Telephone Services         │
│     ATM                                 │
│     Events Routing Discriminator        │
│     Security and Access Control          │
│     IP                                  │
│     SIP                                 │
│     DHCP Configuration                  │
│     ██████████8&9 Series                │
│     Encryption                          │
│     Passive Com. Server                 │
│     SNMP Configuration                  │
│                                         │
└─────────────────────────────────────────┘
```

Getting access, a malicious hacker could launch all the attacks described previously against confidentiality, integrity, and availability. At this point we must note that most systems have a comprehensive on-line help. So, even if the intruder is not perfectly familiar with the system he can be assisted. Figure 3.24 shows the results of a single question mark query "?" at the management prompt. A long list (truncated in Fig. 3.24) is returned with all the commands.

```
<?
CHSEAU : Change session authority class
CHACIV : Change relation authority class index and value
CHSEPR : Change session protection level
CHTRFC : Change traffic class of DNR
DITRFC : Display traffic class of DNR
CHDFCM : Change default facility class mark set for extensions
DIDFCM : Display default facility class mark set for extensions
ASFACM : Assign facility class mark
ERFACM : Erase facility class mark
DIFACM : Display facility class marks
DIFCSU : Display facility class mark summary
SETINS : Set service condition to in-service
SETOUT : Set service condition to out-of-service
SETNIN : Set service condition to not-installed
DISERV : Display service condition
FISERV : Find resources with given service condition
FRCOUT : Force service condition to out-of-service
```

Fig. 3.24   (Partial) list of commands in a command line interface

3.9   Low Level Tools

Besides the everyday tools that administrators use, an array of low level, powerful commands, and tools are available, sometimes non-documented and restricted for highly experienced personnel. They allow to:

- Secretly listen into other connections (by placing a tap) (example in Fig. 3.25)
- Examine memory contents (hex editor) and change then on the fly
- Verify if a line is busy, and if so enable an intrusion
- Send binary commands directly to the CPU or to specific boards
- Enter key presses in the set as if they were entered by its user
- Send direct commands to the set, controlling it remotely (e.g., switch the microphone on)
- Set out of service trunks and sets
- Monitor the signaling in the ISDN lines, in the sets, and in the PBX itself
- Monitor the keys pressed in a set
- Dump the contents of the database holding settings and accounts
- Force connection among two sets (effectively allowing eavesdropping) (example in Fig. 3.25)

There are even cases of forgotten, leftover tools from the testing phase that made it to the production, even with undocumented or hardwired (non-changeable) username/passwords. Special codes and key sequences can also enable hidden functionality.

Debug-maintenance features are very dangerous in the hands of an attacker since they can monitor or isolate single lines or whole trunks. They can also provide a signaling analysis, tracing all messages exchanged. Other tools allow direct access and operations to the database, bypassing the management suite. Memory reading with hex editors provided and hot patching is also possible allowing breakpoints to be inserted in the code and possibly elevate user privileges or bypass passwords. Reprogramming the flash memory can enable secret functionality, something that could be exploited by a malware as we will see later.

Telnet [redacted] pbx
(101)xa00101>
(101)xa00101> [redacted]
- [redacted] : timeslot connexion
(101)xa00101>
(101)xa00101> [redacted]
Thu Jan 26 14:21:11 EET 2012
MANUAL CONNECTION OF EQUIPEMENTS
0 : Exit
1 : One-way connection of 2 equipments
2 : Both-way connection of 2 equipments
3 : Tone connection
4 : Connection from an equipment to a DSP
5 : Connection from a tone to a DSP
6 : Release of equipments
7 : List of active connections
Your choice : 0
(101)xa00101>
(101)xa00101> [redacted]
-[redacted]: for agreement:forces connexion
(101)xa00101>
(101)xa00101>

Fig. 3.25 Commands to multiplex or force connection among two sets (effectively eavesdropping)

3.10 Database

Apart from low level tools, there are tools to access the internal database of the PBX. Indeed, all necessary PBX setup and operation information is stored in such a database that can be accessed either from the management session or (even worse) directly from the O/S (e.g., via open listening TCP sockets). This functionality although dangerous is important for communicating changes in a network of PBXs.

Changes to the settings made via the management suite get reflected in the database. Using the right tools, however, an attacker can bypass the management suite as stated and enter, modify, or delete values directly. This can lead to unexpected results and buffer overflows. Consider the following example: The graphical user interface has a field for a given parameter that allows a single digit to be entered. That value is written in the respective file in the database. An attacker can directly issue a database command to enter into that field a longer entry. This could lead to PBX halting or other problems.

Besides that, having full access the intruder can modify features (e.g., remove call barring and elevate call permissions). He can also delete the whole database. Quite interestingly, at least one manufacturer has specific login accounts with the sole purpose of halting the PBX or deleting and reinstalling the settings database, effectively wiping the existing setup.

```
┌─────────────────────────────────────────────────────────────────────┐
│ ▣ Telnet ▂▂▂▂▂▂▂▂    pbx                                      _ □ ✕ │
├─────────────────────────────────────────────────────────────────────┤
│26/01/12   EDSBR V. 4.00 Page :    1                              ▲  │
│+---------+---+---+---+---------------+----------+------+             │
│┆dir      ┆act┆cpl┆pos┆name           ┆typ       ┆code ┆             │
│+---------+---+---+---+---------------+----------+------+             │
│┆4901     ┆ 2 ┆ 16┆ 0 ┆test test       ┆4012      ┆0000 ┆             │
│┆4903     ┆ 2 ┆ 16┆ 2 ┆test test       ┆4012      ┆0000 ┆             │
│┆4904     ┆ 2 ┆ 16┆ 3 ┆test test       ┆4012      ┆0000 ┆             │
│┆4905     ┆ 2 ┆ 16┆ 4 ┆test test       ┆4012      ┆0000 ┆             │
│┆4906     ┆ 2 ┆ 16┆ 5 ┆test test       ┆4012      ┆0000 ┆             │
│┆4907     ┆ 2 ┆ 16┆ 6 ┆test test       ┆4012      ┆0000 ┆             │
│┆4908     ┆ 2 ┆ 16┆ 7 ┆test test       ┆4012      ┆0000 ┆             │
│┆4909     ┆ 2 ┆ 16┆ 8 ┆test test       ┆4012      ┆0000 ┆             │
│┆4910     ┆ 2 ┆ 16┆ 9 ┆test test       ┆4012      ┆0000 ┆             │
│┆4911     ┆ 2 ┆ 16┆ 10┆test test       ┆4012      ┆0000 ┆             │
│┆4912     ┆ 2 ┆ 16┆ 11┆test test       ┆4012      ┆0000 ┆             │
│┆4913     ┆ 2 ┆ 16┆ 12┆test test       ┆4012      ┆0000 ┆             │
│┆4914     ┆ 2 ┆ 16┆ 13┆test test       ┆4012      ┆0000 ┆             │
│┆4915     ┆ 2 ┆ 16┆ 14┆test test       ┆4012      ┆0000 ┆             │
│┆4916     ┆ 2 ┆ 16┆ 15┆test test       ┆4012      ┆0000 ┆             │
│┆4917     ┆ 2 ┆ 16┆ 16┆test test       ┆4012      ┆0000 ┆             │
│┆4918     ┆ 2 ┆ 16┆ 17┆test test       ┆4012      ┆0000 ┆             │
│+---------+---+---+---+---------------+----------+------+             │
│26/01/12   EDSBR V. 4.00 Page :    2                                 │
│+---------+---+---+---+---------------+----------+------+             │
│┆dir      ┆act┆cpl┆pos┆name           ┆typ       ┆code ┆             │
│+---------+---+---+---+---------------+----------+------+             │
│┆4919     ┆ 2 ┆ 16┆ 18┆test test       ┆4012      ┆0000 ┆             │
│┆4920     ┆ 2 ┆ 16┆ 19┆test test       ┆4012      ┆0000 ┆             │
│┆4921     ┆ 2 ┆ 16┆ 20┆test test       ┆4012      ┆0000 ┆             │
│┆4922     ┆ 2 ┆ 16┆ 21┆test test       ┆4012      ┆0000 ┆             │
│┆4923     ┆ 2 ┆ 16┆ 22┆test test       ┆4012      ┆0000 ┆             │
│┆4924     ┆ 2 ┆ 16┆ 23┆test test       ┆4012      ┆0000 ┆             │
│┆4925     ┆ 2 ┆ 16┆ 24┆test test       ┆4012      ┆0000 ┆             │
│┆4926     ┆ 2 ┆ 16┆ 25┆test test       ┆4012      ┆0000 ┆             │
│┆4927     ┆ 2 ┆ 16┆ 26┆test test       ┆4012      ┆0000 ┆             │
│┆4928     ┆ 2 ┆ 16┆ 27┆test test       ┆4012      ┆0000 ┆             │
│┆4929     ┆ 2 ┆ 16┆ 28┆test test       ┆4012      ┆0000 ┆             │
│┆4930     ┆ 2 ┆ 16┆ 29┆test test       ┆4012      ┆0000 ┆             │
│┆                                                    ┆             ▼  │
│+---------+---+---+---+---------------+----------+------+             │
│◄                                                        ►           │
└─────────────────────────────────────────────────────────────────────┘
```

Fig. 3.26 List of terminal numbers along with their secret codes

Moreover, some kinds of logs, including management logs, access logs, and security logs, are stored in databases instead of plain files, so access to the database allows reading of these files too. Figure 3.26 shows the output of a tool reading the sets database that also displays the secret code of each set (initially set to 0000 by default).

3.11 Non-predicted Feature Interaction

We will now move to features and services. Modern PBXs offer a wealth of services and functionality. There are literally hundreds of services that can be offered, ranging from the simple call forwarding to advanced multimedia teleconferencing. Either they are offered on internal-local level or they are part of network's features depending on the service provider. The interesting point that must be noted here is the possibility of non-predicted feature interaction and the presence of software errors and bugs [4]. As already stated, there is extensive complexity and significant systems interaction. It is therefore logical that a thorough testing of all possible scenarios and parameters that can lead to vulnerabilities is a process that cannot be applied with absolute success.

As an example, let us consider a bug affecting a long chain of forwards. Set A is forwarding to set B, B to C, C to D, and so on. Due to that hypothetical bug, if that long chain of forwards gets back to set A, after having passed through ten different numbers, the system could crash. Another example, if set A has enabled features B, C, and D and if set E asks for service F that involves set A, then another hypothetical case could be that the PBX responds in a non-predicted way, since it was not tested for that specific occurrence.

3.12   The Most Exploited PBX Services

Leaving aside the malicious manipulation of simple features, typical methods of abuse by malicious hackers involve the misuse of more complex and powerful PBX functions such as direct inwards system access (DISA), voicemail, and auto attendant features. In the following sections, we will discuss about the most sought after services that fraudsters are targeting in order to exploit PBXs.

3.12.1   Direct Inwards System Access (DISA)

DISA is designed to allow remote users to access a PBX to place calls (specifically long distance calls) as if they were sitting at their desk in their office. All the user has to do is dial the calling number of the service. This initial call can even be on a 800 free to call line. Depending on the configuration, the system should ask the personal identification number of the user, to authenticate him and grant him access. There is also the option of logging-in into another extension, using both the number of the extension and the respective PIN. A traveling salesman could use this service in order to place business calls while on the road. Fraudsters unfortunately are another category of remote users. Using war dialing they try to pinpoint the DISA service number. Finding systems without codes or with easy to guess ones (e.g., 1234) they proceed to heavily exploiting the PBX for free calls or call selling operations. DISA is most probably the heaviest abused system of PBXs.

3.12.2   Voice Mail

Voice mail use (as well as using a home answering machine) poses three possible threats. One is that if wrongly configured, it can provide access to dial tone in order to place a call. Using a smaller range "war dialing" technique, once inside a voice mail box, an attacker would try to find if there is a code that allows access to an external line. The second one is the inherent dangers of stealing the information contained in them (remember "The News of The World" scandal regarding hacking celebrities' voice mails [5] and the older HP-Compaq merger case [6]) or modifying or deleting (the "News of The World" was accused of deleting evidence from intercepted voice mail). The third one is taking them over. Hacked or specifically created for the purpose mailboxes can be used by illegal ring operations to securely exchange untraceable information.

In addition, these systems are susceptible to denial of service, either by recording lengthy messages that fill up the whole message time duration available for the user, or by inserting DTMF tones in the message itself that command the message to be replayed. This way, a constant loop can be achieved that does not allow the user to proceed with listening to the messages following.

On another note, a very effective trick with a compromised voice mail box is targeting collect call services. With such operator services, a caller can request that the called party is charged for the communication (e.g., it could be a student calling back his parents and asking that they pay for the call). The process is highly standardized and the operators are asking the same questions with the same order and the same timing. Practically, they are asking the name and if the called party accepts the charge. They are also informing him that the cost will be so and so. What intruders do is prerecord all this information in a hacked mail box, as if a living person was replying in the questions of the operator. So a typical recorded message would be "(silence for 5 seconds)… Yes this is me (silence for 10 seconds) Oh, yes I was waiting his call, please connect him through (silence for 10 seconds) yes I realize that I will be charged for the call. Thank you very much." With proper timing chances are that the operator will not tell the difference.

3.13 Complementary Systems

Along with DISA and Voice mail, PBXs use a series of complementary systems and features, including automated attendant (AA), interactive voice response (IVR), automated call distribution (ACD), and computer telephony integration (CTI).

The automated attendant is the system that substitutes a living attendant in order to connect incoming calls to the respective users. It is the system that greets the caller and prompts "Press 1 for sales, 2 for accounting," etc. Attackers go war dialing options (they will try pressing 0 or 9 or other numbers not explicitly stated in the recording). This way they might jump into internal extensions or even worse they might get access to outgoing lines to place calls. It is also possible to find extensions that cannot be called from the public network (non-DDI extensions). The remote management modem is such a case. It might not have been placed under a directly reachable number. Instead, it might have been placed in a specific number to be reached by the automated attendant. In essence, what we see here is a shift of war dialing from the public network to the automated attendant.

Other highly complex systems that can be manipulated include the IVR. This is a more advanced system that allows the user to actually say his option rather than press a key. So the system asks the caller, "Say sales to get connected to sales, Accounting to get connected to the accounting office," etc. There is also the ACD system that automatically distributes incoming calls to a specific group of terminals that agents use. It is interesting to note that, in many cases, AA, IVR, and ACD are separate systems, based on software and/or infrastructure provided by 3rd parties, connecting to the PBX; as such they have their own vulnerabilities. In any case, a closer connection with the IT infrastructure, with VoIP and computer telephony integration (CTI) systems opens the door for an array of attacks originating from the computer network and the other way around.

3.14   Other Dangerous Points

The reader should bear in mind that attacks, not directly aimed at the PBX but rather to its surrounding infrastructure, can have an equally critical impact. A range of technologies complementary to the main PBX functionality can affect its security. There are various wireless connections between sites (e.g., microwave or infrared). DECT infrastructure for wireless phones is possibly installed. PBXs can also be connected with fixed cellular terminal (FCT) that bridge calls to mobile phones with the network of the mobile phone operator, offering cheaper calls. Electromagnetic frequencies can cause interference or leakage of information. Due to lack of space we will not extend the analysis, in order to focus on VoIP.

3.15   On VoIP Security

Once again, Voice over IP (VoIP) deserves a section on its own. VoIP commonly refers to the communication protocols, technologies, methodologies, and transmission techniques involved in the delivery of voice communications and multimedia sessions over Internet Protocol (IP) networks, such as the Internet [7]. With the introduction of these technologies the situation gets even more complicated. Telecommunication administrators need to become even more alerted to prevent new and existing threats that are typically associated with data networks, now impacting voice networks, and vice versa.

Conventional PBXs typically use proprietary protocols and specialized software and have fewer points of access than VoIP systems. With VoIP, opportunities for eavesdroppers and fraudsters are multiplied [8]. Without the necessary security measures in place telecoms systems can become the weak link in the network. Attacking them can lead to the compromise of the IP network of the company, with the PBX providing the "back-door" to enter a secure (nonetheless) network.

VoIP suffers from a number of dangers that are either unique to it or manifest as different forms of the dangers classical PBXs face. References [9–11] provide a comprehensive overview of these dangers. We will try to give an overview in the following paragraphs.

There is extreme complexity of features and openness and modularity in independent implementations and products comes with its own vulnerabilities. Therefore, developers overwhelmed by the complexity of protocols might ignore details that are crucial for the security of the protocol exchange. Implementation faults and bugs are manifesting with empty, malformed, or large volumes of signaling messages. At the same time, protocol responses to carefully crafted messages can reveal information about the system or its users to an attacker.

VoIP is implemented on top of IP infrastructure, including protocols and services such as DHCP, DNS, RADIUS, TFTP, BOOTP, NAT, STUN, NTP, SNMP, HTTP, TLS/SSL, Routing protocols, and others. Securing the VoIP overall means the administrator has to always keep secure all these supporting protocols. Failing to do so can critically affect VoIP. As an example, lack of proper authentication by the registrar or proxy, or by the SNMP server can lead to traffic interception and user impersonation. Other such examples include insecure wireless environments, enabling man-in-the middle and eavesdropping attacks.

Since VoIP devices are primarily software based, they are vulnerable in every aspect generic software is susceptible to (e.g., buffer overflows from bad input validation). The integration of side functionality in VoIP such as a web server for the management interface allows attackers to take advantage of vulnerabilities found in the surrounding systems and not in VoIP per se. Specifically for web services, their cross-interactions make attacks such as cross-site scripting (XSS), SQL injection, and cross-site request forgery (CSRF) effective. It can even lead to privilege escalation. In addition to that, code reuse and shared software in order for other systems to interact with VoIP systems lead to spread of the bugs, affecting other platforms too.

Session initiation protocol (SIP) itself, which is the fundamental protocol behind VoIP, has a large complexity with known vulnerabilities, leading to confidentiality and availability problems. Nonrandom enough or predictable values in SIP header messages as well as specially encoded uniform resource indicators (URIs) can be a source of problems. Other protocol-level vulnerabilities also exist, possibly because of bugs or unforeseen interaction paths.

Furthermore, if registration requests are not properly handled, then attackers may be able to receive messages intended for other users. Various attacks in the Digest Authentication (such as actual URI and SIP INVITE differentiation) allow for credential reuse and toll fraud with relay attacks. In these attacks the fraudster can reuse another party's credentials to obtain unauthorized access to the services.

Technically, VoIP systems can particularly be abused to spoof the caller ID. This is possible both because of the more decentralized and free nature of Internet (compared to Telephony networks) and because of the lack of cross-checking of information across several messages during call setup and throughout the session.

As is the case with traditional PBXs, undocumented commands and features left enabled by default are a serious cause of problems. Default configurations and login and passwords are as much a problem in VoIP as in every other technology with lists of default accounts easy to find on the Internet [3].

3.16   On a PBX Malware

VoIP evolution was a major step towards the modernization of traditional PBXs that were closed systems with proprietary operating systems. Many of them are now UNIX–Linux based, even using open source components. In these operating systems developers can write code the same way they do with computers. Besides the increased usability, the more complex a system is the more security gaps it has. Along with legitimate applications, viruses and other malicious programs can emerge.

It is well possible to implement a malware effectively targeting PBXs. Such an occurrence would have a devastating effect on the communication confidentiality, integrity, and availability. This effect can further be amplified in warlike situations. At the same time, the close connection and convergence of telephone and IT infrastructure can lead to PBX disruptions having a broader impact. PBX malware could take down telecommunications, intercept sensitive calls, harvest data from call logs, and lead to performing psyops during war. It can also be used to support economic fraud and money laundering and has the potential to lead to millions of Dollars of damage per day, critically affecting the economic well-being of any company or organization. The possible effectiveness of malware targeting critical infrastructure has been proved on world scale by Stuxnet [12, 13].

Along the lines of the development of PC viruses, PBX malware could follow a comparable development. Viruses first just flashed messages, then erased hard discs—and now steal information or insert erroneous information. PBX malware could take down telecommunications, intercept sensitive calls, harvest data from call logs, and also lead to performing psyops during war.

Such a malware can be implemented using an array of tools and functionalities modern PBXs offer. Even before the VoIP era, there existed PBXs that used a core operating system based on UNIX, offering a fully enabled programming environment with shell command functionality, time programmed batch jobs and cron tables and programming language compilers such as gcc. For those PBXs that do not utilize such an operating system, there is usually some form of proprietary scripting or interpreted language for the specific platform. Finally, even if they are absent, automation and testing tools such as *Expect* [14] can always be used in the attacking node, effectively controlling the PBX under attack, as if the code was executing in the target. In a similar way, many terminal emulators too have their own scripting language that can be used to command the target PBX from the attacking PBX session.

In any case, the code or script used by the malware would target the management of the PBX, effectively issuing commands or changing settings. Such changes could take place using the standard management platform, or even better using lower level commands and tools. Indeed, most PBXs have a list of commands and tools (sometimes not documented) that can provide access to the interworking of the switch and achieve almost everything as we previously described.

The malware's algorithmic block is presented in Fig. 3.27. It is looking for targets to infect, verifies they are indeed PBXs, penetrates their security measures, uploads in their operating system the corresponding compatible version, depending on the

Fig. 3.27 Algorithmic
block of malware

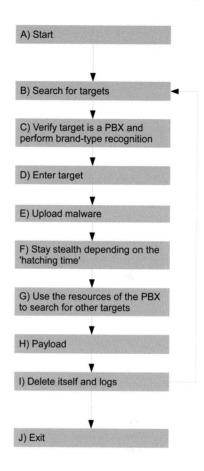

actual PBX brand and type, stays stealth for as long as it is programmed to, activates
its payload, looks for other PBXs to move to, and finally erases the logs and exits.

More analytically, the steps are given in the following paragraphs.

3.16.1   Start

The launch of the malware can take place from a compromised PC (preferably
belonging to a company owning a PBX), or from an already manually compromised
PBX. Given the convergence of voice and data infrastructure, the malware could
involve computer networks spreading its functionality too. In that case, compromised
computers actively participate in the search for other PBXs and new PCs to infect. It
is also possible to manually plant the malware either in software or hardware with
physical access to the PBX, even combining some form of social engineering. Due to
lack of space we will focus only on software PBX infection and spreading.

3.16.2   Search for Targets

At this step, the malware will try to find information about other PBXs in close relation to the already infected host. To help the automated data gathering step, the malware could possibly include an initial list of targets, manually compiled from yellow pages, to facilitate the spreading. The list could contain the published numbers of ministries, banks, industrial facilities, companies, etc., so that the malware could try war-dialing on them. In any case a tracking mechanism should be deployed, possibly via a command and control center, so that the same PBXs are not attacked twice.

3.16.3   Verify the Target Is a PBX

Once a connected system is found, the malware would assess if it is a PBX. Most PBXs have distinctive login prompts or error messages which makes accessing the brand type a trivial task. In case the system cannot be recognized, it could be logged for further manual action. As already stated, a more evolved malware could incorporate functionality to spread to IT platforms too, but we will not cover such an option here.

3.16.4   Enter-Break into the Target

Using default passwords or known vulnerabilities the malware would enter the target. The database of passwords the malware maintains would be gradually extended, since it will also have the ability to harvest further passwords from the attacked systems during its spreading. In the unlikely event that none of the default passwords works the malware could attack using known vulnerabilities for the specific platform. This option heavily depends on the platform of the host PBX that is launching the attack. Indeed, in order to use specific tools and commands, the host PBX should be using a general purpose operating system. This is the case not only for VoIP solutions but also for some traditional PBXs. Should all means fail, the malware would abandon the break-in process and return to the previous step, picking another target.

3.16.5   Upload Itself and the Payload

At this step the malware uploads its propagation vector and the payload to the compromised PBX. They are both specific for each brand-type, but the whole "package" for all possible brand types could be uploaded in library files, in order to be transferred to further targets, connected to the already compromised PBX. It is usually the case that two to three specific brands cater for the largest percentage of PBXs in most countries, depending on vendors' market share. Malware could be tailored to

proceed with only the most usual platforms per country (or even in a single platform), thus being more compactly sized and more effective.

3.16.6   Stay Stealth Until the Period of Activation (Hatch Period)

According to a well-known virus technology, the payload is usually not immediately activated, in order to facilitate the spreading of the malware. Indeed, should
the malware immediately manifest, administrators could possibly detect and remove
it before having the chance to spread to more systems. On the other hand, a very
long period of inactivity could result to the malware being detected in a routine
check. For the specific malware, the PBX scanning (not including the war dialing
option) and data harvesting could start early, since they are noninvasive. In any case,
when enough new targets are found and compromised, the actual payload could
activate itself. A command and control center could again supervise the process.

3.16.7   Use the Resources Compromised to Find Other PBXs

After the inactive period of the previous step, scanning for new targets will start.
Initially this could include low profile activity, such as data gathering from files present
in the O/S of the PBX. Subsequently, the active outdialing to targets would start, during
the night and holidays as to remain as much as possible unnoticed. The whole process
iterates back to the looking for targets step, before either a time trigger or a command
from the control center forces the malware to activate its payload as follows:

3.16.8   Activate the Payload

The "payload" of the malware is the part of it that performs the malicious action. In
the case of the PBX malware we are discussing that payload can have different
forms, affecting confidentiality, integrity, and availability as we have seen throughout the book.

3.16.9   Delete Itself and Logs

Having already infiltrated new targets and having activated its payload in the current
target, the malware would try to cover its track by deleting itself and the logs associated with its actions. It could even proceed to full wiping of the O/S of the target. It
must be noted at this point that records in call logs are registered after a call ends,

since they have to contain information about the duration of the call. In the case where the malware has propagated with a modem dial up call, it is impossible to delete the record that points to the number of the previous PBX that infected the present one. This is particularly important in low end PBXs that do not offer advanced O/S functionality such as time scheduled jobs (like UNIX cron tables). In such PBXs the malware has to call the target many times, each time deleting the logs associated with its previous functionality. Still, however, the last call will remain logged, unless the malware proceeds to a full wipe and crash-shut down of the PBX. As a side note, this paragraph reinforces the importance of having external logging platforms, independent of the PBX itself that regularly can be backed up.

3.17 Conclusion

This chapter has been the most technical chapter of the book, so far, covering as much as possible the components that PBXs consist of, in software and hardware level. Having provided the technical background and the way services get abused, in the following chapter we will focus on securing PBXs.

References

1. Androulidakis I (2009) On the importance of securing telephony systems. Wseas Trans Commun 8(1):102–111
2. Baker WH, Wallace L (2007) Is information security under control? IEEE Secur Priv 5(1):36–44
3. Virus.org (2012) Default password. http://www.virus.org/default-password/
4. NIST (2001) PBX vulnerability analysis. Special publication 800–24, 2001
5. Wikipedia (2011) News international phone hacking scandal. http://en.wikipedia.org/wiki/News_of_the_World_phone_hacking_scandal
6. Computerworld. HP Exec feels violated by voice mail leak. http://www.computerworld.com/s/article/70061/HP_exec_feels_violated_by_voice_mail_leak_
7. Wikipedia. Voice over IP. http://en.wikipedia.org/wiki/Voice_over_IP
8. Walsh TJ, Kuhn DR (2005) Challenges in securing voice over IP. IEEE Secur Priv 3(3):44–49
9. Keromytis AD (2009) A survey of voice over IP security research. In: Proceedings of the 5th international conference on information systems security (ICISS), Kolkata, India, pp 1–17
10. Keromytis AD (2011) Voice over IP security, a comprehensive survey of vulnerabilities and academic research. SpringerBriefs in computer science, 1st edn. Springer, New York, XIII, 83 p
11. Keromytis AD (2009) Voice over IP: risks, threats and vulnerabilities. In: Proceedings (electronic) of the cyber infrastructure protection (CIP) conference, New York, NY, June 2009
12. Chen T, Abu-Nimeh S (2011) Lessons from Stuxnet. Computer 44(4):91–93
13. Langner R (2011) The first deployed cyber weapon in history: Stuxnet's architecture and implications. In: CCD COE international conference on cyber conflict, Tallinn, Estonia
14. Libes D (1991) Expect: scripts for controlling interactive processes. University of California Press, Berkeley, CA

Chapter 4
PBX Security

4.1 Introduction

While data communications have long before begun to utilize every possible means of protection, enjoying a vivid research and development sector, PBX arena has not caught up. Following the taxonomy of threats and the technical analysis of the previous chapters, in this chapter we will further provide some useful tips and advices for safeguarding PBXs. Due to the multitude of dangers modern PBXs face, centralized actions are needed in order to both educate the users and secure their telephony systems. Such projects can consist of educational, policy, auditing, technical, documentation, hardware and software solutions and actions, as will be described in the following paragraphs.

4.2 Physical Security

Before proceeding to the actual PBX security measures, we will discuss physical security. In the physical domain, equipment and administration sites should be closely controlled and monitored. Access control measures including locked doors and locked racks, area surveillance with CCTV and alarms, coupled with environmental monitoring (temperature, humidity levels), and fire and flood protection can reinforce the physical security. The sites where this infrastructure resides should not be advertised with signs and other informatory posts and labels. They could as well be "hidden." It is one of the rare occurrences where security through obscurity is actually a good tactic (contrary to encryption algorithms where a public design subject to public scrutiny is the best way to go).

Protection from unauthorized access is essential since access to the PBX premises allows for a "bug" or other interception device to be planted there, literally at the heart of the telecommunication network. At this point, we must note that specific

© Springer International Publishing Switzerland 2016
I.I. Androulidakis, *VoIP and PBX Security and Forensics*, SpringerBriefs
in Electrical and Computer Engineering, DOI 10.1007/978-3-319-29721-7_4

care must be given to avoid the dangers associated with cleaners and janitors that might have access to the site.

Another usually forgotten aspect is the protection against environmental elements and disasters. A fire could burn the infrastructure endangering human lives too. A flood could prove extremely harmful for the sensitive and expensive equipment, while a water pipe leak can cause a severe damage and a complete collapse of the network which will not be easy to deal with. It is of great importance to take all the appropriate measures to guard against such incidents. In case of a natural disaster, such as an earthquake, there should always be provision for disaster recovery procedures that we will discuss in section. Moreover, keeping the place clean and tidy can always help the technicians do their work, especially when time is critical during an incident. Tidiness eases the frequent optical examination of equipment and cabling for interventions in the hardware and the lookout for new/unknown/unlisted equipment (e.g., a new modem).

Finally, part of the physical security involves public phones such as the ones found in elevators or conference rooms. They should have their calling capabilities minimized. Furthermore, the operator's console which is also a set that can be accessed during off-hours must be properly secured, by disabling administration features and using authentication features with the operator always logging off when leaving. As an example, the command to disable programming features from the operator's console for a given brand is as simple as this: OPCAC:DIR = 1000, PRG = 0, where it is assumed that the operator's call number is 1000. Although simple, if this step is not taken, then as already discussed in Sect. 3.6 the operator's console maintains the ability to change PBXs setup features and operational data of other extensions. That could easily be exploited by an attacker.

Apart from public phones, analog cordless phones should be banned, since they offer no protection against eavesdropping. DECT cordless phones are generally encrypting communication but should be checked against the tools discussed earlier, to make sure that they are not vulnerable [1]. In any case, they are susceptible to jamming attacks, so they should not be used in critical operations.

4.3   Nontechnical Security Issues

Following physical security, no matter what technical security means might be in place, nontechnical measures play an equally important role. Human resources are the most valuable assets of a company and critically affect the overall security level. Education and awareness are therefore always needed. Specific user categories need the respective training and educational programs of varying length, according to their involvement with the PBX. Starting with plain users, they should be informed about the importance of passwords and PINs. Administrators need extensive technical training. Telephone operators are usually the ones who mostly come across social engineering, so they should be trained to avoid being a victim of such an attack. Technicians, finally, should be trained to report unusual interventions or

VOICE MAIL WORK INSTRUCTION	**Document Control** Reference: DOC 5.3 Issue No: 1 Issue Date: 10/01/2012 Page: 1 of 1

1   Scope

All Organizational voicemail services are subject to this instruction,

2   Responsibilities

All employees, sub-contractors or temporary staff who have a voicemail facility are subect to this instruction. The telecommunications Manager is responsble for the configurations of the Organizational telecoms services

3   Work Instruction

3.1   Voice mail messages must not provide information that might enable a potertial miscreant to identify the perion or nature of the voicemail owner's ansence (such as being on holiday). They must simply say that the recipient is not available, provide an afternative contact, and ask for a name and number to be left so that the call can be returend.

3.2   voicemail messages may not include any information that might be classified as either confidential or restricted.

3.3   The voicemail system has s standared statement to callers that tells thern not to leave confidential information on the voicemail system.

3.4   voicemail passwords must be individuatized and changed on a regular basis to reduce the possibility of unauthorized messages being created.

Fig. 4.1 Voice mail work instruction

changes in the equipment and the cabling. Standard Operating Procedures, Security Manuals, and Users' security police guides can enhance the educational programs.

Formal security procedures and policies can also help mitigate the danger of social engineering. Users should be informed about the allowed and approved use of telephones and PBX features as stated in the relevant policies. They should also sign the respective documents stating their agreement to the terms. It is of paramount importance that procedures are properly communicated in order to actually be followed. An information security management system (ISMS) such as ISO 27001 can provide the necessary framework for the proper design of nontechnical security measures. As an example, Fig. 4.1 shows an excerpt from a work instruction regarding the use of voice mail [2].

Common sense and tidiness can help a lot. The importance of properly and tidy documentation of equipment, cabling and patching and connections status manifests not only during a malfunction but also when tracing an incident. Technicians can perform their job more effectively and in a shorter time. At the same time the documentation makes it possible to easily identify and remove any "external" elements, such as "bugs." Moreover, in case of a disaster, or other major malfunction, proper labeling and documentation could speed up the repair time. A simple in-house cabling documentation system that registers not only patching data but also user identification data is an effective and easy to implement solution.

PBX should be incorporated into the disaster recovery and business continuity plans. There have been many cases in the past, where the PBX was not included in the disaster recovery plans! IT service was restored but the company left without phones could not operate anyway. Specific incident response procedures should exist, following the principles that we will analyze in the Forensics chapter. Along the same line, insurance programs should be assessed to see if a proper and convenient insurance option is available. In case a company is collocated in a building with another company, mutual agreements could help in cases of crisis, where some basic telephone service can be provided by the infrastructure of the other PBX. Complemented with telephony disaster recovery plans that include backups, reserve trunks, and cabling, this step would cover among other things part of the requirements of an ISMS such as ISO 27001.

Speaking of disaster recovery, backups should be thorough and frequent. They should include operational and administrational data. Ensuring they actually work is also important since there are many cases where a backup has been corrupted and failed to restore the system. Given that many PBXs use proprietary O/S, the backup should be accompanied by the respective tools to reinstall it. In any case, the backup archives should be encrypted to avoid the unpleasant case of theft of backups that effectively hands all the contents to the thieves.

Protecting the equipment is not enough. Information regarding the PBX setup itself must be protected. Documentation copies and manuals should be kept in a secure place. Attacks can be facilitated if a press release disclosed the buying of a new PBX with all the features installed and the users it will be serving. Although tempting to advertise a new, state-of-the-art network of PBXs full of new functionality and modern technologies, at the same time it is dangerous to become a welcome call for fraudsters that would immediately target it. The author still remembers a certain company offering call center services, advertising in its webpage that its PBX has DISA service! (Remember that DISA is the mostly abused by fraudsters PBX feature). In addition to that, it is never a good idea to publish the full catalogue-numbering plan of the company in the Internet. Name search limited to a few queries per user, or general reception only numbers should be provided in the yellow pages and the web pages of companies. Publishing the whole catalogue helps not only social engineering attacks (since names and business capacities of employees are communicated) but also war dialing (since fraudsters know which numbers are voice numbers and can restrict their war dialing in the numbers not presented in the catalogue).

Even the company trash needs protection and proper ways of disposal. Hackers that have targeted a given PBX network often use the so-called "dumpster diving" technique. The technique is carried out by just inspecting the company's trash hoping to find valuable data. Such data can give them valuable information about security protocols, anti-hacking measures, the topology of the network, and possible soft spots in security or in the infrastructure. Even worse, they can provide access codes and usernames which can lead to direct network exploitation. It is therefore of great importance to destroy all sensitive data before disposal. Paper shredders and low level formatting of HDD can help in that. Similarly, manuals, directories, and other internal documents should be treated as confidential.

Table 4.1 List of nontechnical security measures

Enforce formal procedures and policies
Educate users, administrators, and operators
Keep everything documented and tidy (equipment, cabling, connections, setup) Perform frequent optical checks
Incorporate PBX into DRP and BCP plans
Take frequent, encrypted backups and test they work
Check insurance programs
Have mutual agreements with other companies for collocation/equipment sharing services
Implement incident response procedures
Keep contact with installer–provider–manufacturer
Stay updated about the dangers
Use paper shredders (against dumpster diving)
Do not advertise the company's full catalogue-numbering plan and the PBX itself
Consider implementing an information security management system such as ISO 27001

Since PBXs are evolving, there must be a continuous collaboration with the provider and the distributor/manufacturer with frequent briefings on the security of the product. This way updates and patches will be made available and of course installed as soon as they are published. In any case, before installing anything new, it should be checked with the manufacturer too.

Closing the section, Table 4.1 synopsizes the nontechnical security measures we described so far.

4.4 Technical Security Issues

Although nontechnical security means are essential, when dealing with highly advanced technological pieces of equipment, such as PBXs, technical measures are unavoidable. In the following sections we will shed some light on the issue.

4.4.1 Local and Remote Management

PBXs should preferably be administered by local teams. This minimizes the need for a remote connection abuse. Access to the management suite through a local console, the local network, or remotely (via dialup lines or IP connectivity) should be secured. Every serial port connection should be traced to its destination.

In addition, the remote access port/feature should be deactivated (switching off the modem is not enough since somebody could switch it back on). When remote access is truly needed (e.g., upgrade from the provider or solving a difficult problem) then it could be activated, following a specific procedure and a defined time schedule. A telephone call could precede where the caller is properly authenticated to secure that he is actually calling for a legitimate purpose. Following that, the port could be enabled, but only for as long it takes for the task to be completed. It should immediately thereafter be blocked again.

In case a modem is used, the dial-in number should be unlisted, preferably in a different numbering plan (e.g., if the company owns the numbering plan 555-0XXX then the modem could be behind number 555-1YYY or even better in 556-ZZZZ). This way, a war dialing attack would fail to find the number. Callback and caller ID options should be used. With the first one, the modem calls back a number already programmed in memory to avoid the risk of a connection from another party. With caller ID checking, the identity of the caller is checked against a database, but this feature should not be the only one to trust since caller ID spoofing is also possible. Moreover, the maintenance modem could be programmed to answer after six or more rings. Most war dialing programs limit the rings to three to four in order to scan as many as possible numbers in the shorter time. Such a long time to answer could help avoid being detected. The idea to map the modem number behind the automated attendant (AA) although better than having it in a plain DDI number is not bulletproof since as we discussed, all it takes is war-dialing the AA instead of the public network.

Specifically for the local management, IT security and network security are essential. Apart from strong passwords to login to the management console, there should be automatic logoff measures and password protected screen savers. A time out-enabled screensaver is important even if the management terminal is situated in a closed and controlled area. This is because using social engineering tricks an attacker could gain access in the premises it is placed. The management software/ suite should also be patched and always updated. Processes running in the administration PC or server should be closely monitored and minimized. Following network security best practices it is important to have segregation of network segments, making sure that the management traffic is not seen by other segments. We will not extend much since these are topics that better fit in the context of computers and networks security literature.

In any case, all PBX accounts should have a login and password, with automatic logoff implemented in cases of inactivity. Needless to say that default passwords should immediately be changed. A measure that can help here is to have password aging where using an automated mechanism passwords must be changed in frequent time intervals. Furthermore, different accounts must exist with progressively more capabilities as to avoid user super-user accounts for everyday trivial tasks. Having different accounts helps for accountability and non-repudiation too during an incidence response procedure. Another technical solution stemming from the banking experience uses a strong, multifactor authentication (e.g., two factor authentication where the user has also to enter a random number produced by a token device). So apart from the password the user has also to respond to the challenge question, as seen in Fig. 4.2. Besides login and password combinations, numerical PINs and codes used for PBX services and tools should be treated as passwords too, with frequent changes, avoiding obvious numbers such as 1111, 1234, 0000, 9999, etc.

Fig. 4.2 Login and
challenge procedure

```
CONNECT 38400

Login:
Login: ███
Password:
INCORRECT LOGIN

Login: ████
Password:
INCORRECT LOGIN

Login: ████
Challenge: 5█████63        Product ID: 1████████

Response:
INCORRECT LOGIN
```

SUSIP:DIR=4004;

STATUS INFORMATION AT 11:24:45 22FEB99

DIR	TYPE	TRAFFIC STATE/	PTR	LINE STATE/	PTR	DIV STATE	ADD INFO
4004	ATS	IDLE	#0095	FREE	#0175	BSY,NAN,ECF	DND ACT

END

Fig. 4.3 A set with external forwarding

4.4.2   Settings and Configuration

Having secured the PBX management console and the PBX management process, the actual settings have to be revised. Call capabilities and permissions should be restricted only to destinations needed. Premium rate destinations should explicitly be banned as well as international calls. Each set should belong to a class of service/ facilities allowance with the minimum amount of services needed. More services should be reserved for only a few sets. Connection classes can also be enforced to prevent certain sets from calling other sets, or receiving calls. Call forwarding to external destinations should be kept to a minimum or barred altogether. As an example Fig. 4.3 shows a set that has external forwarding active (indication ECF).

Checking trunks' occupation and load can give an early sign about malfunctions or possible fraud taking place. This is important not only for the central office lines but also for the trunks interconnecting various PBX segments. An important part is the examination of logs that we will examine in Sect. 4.9. In any case, apart from call logs, changes logs should be kept, pointing also to the specific user account that requested the changes, to help when dealing with incident response and forensics. We will discuss more about management logs in the next chapter.

PBX networks offer a wealth of applications and functionalities that the average user never gets to use. Nonetheless, they are affecting the security of the phone, either positively or negatively. The reader should bear in mind that the less he knows about the features, the less protected he is against social engineering attacks. This is why dangerous services that are installed and enabled willingly or not should be located. Such services include remote maintenance feature, DISA, intrusion, teleconference, set substitution function, forwarding settings, and diagnostic tools state. Should there any embedded or linked automated attendant, IVR system, or voice mail are installed, they should also be checked. Especially in cases where the system is offered by a third party manufacturer, interconnection vulnerabilities should be examined. The most dangerous features such as DISA and voicemail deserve special attention that we will discuss later on in this chapter. It is better to be disabled or even removed if they are of no use. Otherwise, in collaboration of the manufacturer, every suggested measure, patch, and upgrade should be applied. We will see more details later on this chapter. Not only features but commands and tools that can be abused should be disabled or strongly monitored. Even if disabled, they should frequently be checked to make sure they remain disabled, since it is possible that an administrator temporarily allows a feature and then forgets to disable it.

Another important security measure is to disable any tones that prompt the user to enter the PIN. Automated war dialing tools are actually looking for the presence of tones to classify the numbers that are answering. A voice or silent prompt would be more difficult to be automatically classified and probably would be skipped by war dialing software.

4.4.3   Software and Hardware

To help the administrators in their everyday tasks, special software can be offered, specifically tailored to their needs, including management and monitoring features, troubleshooting options, call detail logging as well as a log analysis and data mining functionality. Apart from software, hardware including special PBX crypto devices can also help shield the PBX. Voice encrypting products work by transforming voice to data and encrypting the data before leaving the phone. As such the encryption is not depending any more on the provider–carrier but rather on the specific software and/or hardware suite installed in the phone. Needless to say that the recipient of the call must have a compatible product in his phone too.

Besides crypto devices, in hardware level there are also PBX firewalls available [3]. During the past few years an analogy of computer firewalls has made it into the telecommunication world. These PBX firewalls are connected between the PBX and the provider–carrier and effectively control the parameters of each call. Should a call deviate from the normal call pattern (being too long, or headed towards a new destination, or taking place out of business hours, etc.), then it is logged or disconnected. Moreover, such a solution hosted in an isolated system solves the problem of breaching its integrity. The arrangement of this system is shown in Fig. 4.4.

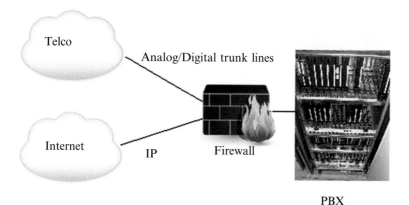

Fig. 4.4 PBX firewall

4.4.4   Audits

After securing the PBX, its overall security level can be assessed with audits. The first step in assessing it would be to perform an external penetration testing, without any prior knowledge of the system, exactly as an outsider would do. Usually this task is delegated to security consulting companies. Using war dialing tools and techniques, modems and other service and feature numbers can be detected. Following, there should be attempts to intrude into the system via any vulnerable entry points found. The penetration test should also be supplemented with attempts to circumvent the security measures from the inside perimeter, as an internal user or as a guest would do. In both scenarios, social engineering attacks could also be tried.

The audit can also include the physical manual examination of distribution frames and other dangerous points for bugs. The documentation of infrastructure can be one of the deliverables of such an audit, containing physical (trunks, wires, MDF status) and logical (administrative settings details) of the PBX.

Given access to the management console of the PBX, a thorough audit should take place, testing the security status of the installation and whether it is correctly implemented. This will be achieved by spotting insecure configuration settings and options that contribute to possible security shortcomings, as discussed throughout the book. In a more advanced level, CRC hash checks in the binaries in the PROMs, firmware, and software could take place to validate they are not tampered with. In the same wavelength, searching memory and HDD for unknown programs–tools can uncover problematic situations.

4.4.5   In Conclusion

Closing the section, in Table 4.2 we present a list with the technical security measures.

Table 4.2 List of technical security measures

Local and remote management
Change default passwords with strong ones and keep changing them frequently
Keep all service security codes (PINs, etc.) secret and change them frequently
Use strong authentication (Password + {smart card, one-time tokens, callback, etc.}) Disable remote management, use it only with callback and specific procedures
Use network segregation
Check running processes in the management O/S Install updates and patches
Settings and configuration
Check settings and configuration
Disable not needed commands/options/features
Check admin accounts and user privileges
Restrict international and premium rate numbers calling capability
Enable logging and check call and administration logs Check with manufacturer before installing anything new Disable DISA or restrict to local calls only
Pay attention to voice mail usage
Configure properly the automated attendant
Do not forget backups (keep them encrypted)
Software and hardware
Consider a PBX firewall
Consider the use of hardware cryptophones or software suites to encrypt communication
Audits
Perform internal and external penetration testings
Ban analog cordless phones
Test any DECT cordless phones used against hacking tools
Document equipment, cabling, and connections
Search memory and HDDs for unknown programs–tools
Perform hash checks in the binaries

We will move onto examining the most abused PBX services, namely DISA, voice mail, and automated attendant.

4.5 Direct Inwards System Access (DISA) Security

DISA as mentioned previously is one of the mostly abused services. It should be better to disable DISA altogether and even uninstall the service from the system, since a malicious hacker could re-enable it if just disabled. Other options such as prepaid calling cards or prepaid mobile phone cards could be considered as an alternative. A written statement from the supplier that the feature is not installed in the system and cannot be enabled could be helpful to avoid cases such as the "Athens Affair" [4]. In that scandal, somebody managed to activate the preinstalled, but not activated (since it was not paid for) system of lawful interception. With a line of highly sophisticated tricks and hacks, he operated the system stealthily intercepting the communications of hundreds of important persons, including the prime minister's. Albeit the attack was launched against a mobile telephony network, the infrastructure targeted was mobile switching centers, which are more or less modified central office switching centers. These in turn are "big" PBXs. As such the incident remains relevant to PBXs.

If DISA is truly needed, it should be securely configured to always request at least a PIN. Even better, different login-pin combinations should be issued for each user to facilitate the accountability and non-repudiation. The dial-in number should be kept unlisted and classified, treated as a password. It could be worth to restrict DISA to local calls only, although that would probably make it redundant as a service anyway. Apart from PIN, caller ID authentication can be used. This way, incoming calls can get through only from designated numbers (e.g., the employees' home phone number). Concurrent calls using the same PIN should be blocked, as this is a clear sign of PIN leakage. Finally, administrators should regularly monitor its status (enabled/disabled) as well as the number of PINs issued/in use. Disabling tones, as suggested in Sect. 4.4.2, is also a very practical measure.

4.6 Voice Mail Security

After DISA, the second mostly abused system is voice mail. First of all, access to trunk lines and forwarding from mail boxes should be disabled. If not, then voice mail essentially transforms itself to DISA, with all the associated dangers. Empty or unused mail boxes can be taken over by attackers, so they should be removed as soon as possible. Moreover, users should be disconnected following repeated wrong passwords or repeated entry of options that do not exist (sign of war dialing the options of the voice mail to get an external line). Even better, such call occurrences could be transferred to the operator, thus alarming him. Users themselves should be advised to use strong password codes, often changed and kept secret. There are options to enforce password changing in regular intervals. Finally, access and usage logs should be monitored and audited for inappropriate use.

4.7 Automated Attendant Security

The third important system, the automated attendant, should also be carefully setup. It should be made absolutely clear that no external destinations can be called. There are many cases of older systems where a single "0" or "9" (options that were not mapped to specific extensions–destinations in the automated attendant rules) would connect the user to an external line, allowing him to call. All unused options in the tree of options should be routed to the operator; the same should happen with repeated mistakes or wrong choices, since they are all signs of war dialing. Routing the calls to the operator instead of disconnecting them can thwart attackers and alert the operator in order to inform the administrator.

4.8   VoIP Security

Returning to VoIP we will provide a short list of essential things to keep in mind. Table 4.3 synopsizes them.

We will now focus onto hardening Asterisk PBXs.

Asterisk, named from the asterisk symbol "*" found in telephone keypads, implements the functionality and the services of a PBX, allowing telephones to make calls, and to interconnect to the public switched telephone network as well as to IP telephony networks. Thanks to its software nature and to the GNU General Public License (GPL)-free software license, it offers unparalleled customization and add-ons that end customers can enjoy. Indeed, users can build telephone systems, add features in already existing networks, or replace existing PBXs.

At the same time, it is released with a second proprietary software license that allows licensees to distribute proprietary, unpublished system components. Although originally designed in the late 1990s for Linux, by now it can be deployed on many other Operating Systems. What is more important is that due to its compact code size it is possible to run in an embedded system, while it can also boot from a flash drive, live CD, or external disk.

Taking advantage of the dual license mode, Asterisk has been used in many products-projects, both open source and commercial ones. Such products are PBX in a Flash, LinuxMCE, FreePBX, Elastix, AskoziaPBX, and others. Many third party commercial products extend its functionality in any imaginable way.

Asterisk's functionality can be extended either using its own extension languages or by adding modules in C. Moreover the Asterisk Gateway Interface allows communication via standard system streams or TCP sockets.

Regarding the protocols supported, Asterisk supports both Voice over IP (MGCP, H323, SIP, etc) and classical signaling both via analog (FXO) cards or digital (T1-E1) TDM ones. Specifically for inter-Asterisk connectivity there also exists a dedicated protocol, Inter-Asterisk eXchange (IAX), RFC 5456.

Table 4.3   Essential VoIP security tasks

Have frequent updates, not only for servers but also for VoIP hardphones
Pay attention to all services running along with VoIP, both in servers and hardphones
Stop or block any unnecessary services
With the help of an expert, experiment with SIP fuzzing tools Harden and protect all VoIP servers following IT security measures Harden and protect the infrastructure on which VoIP services rely
Consider using a redundant server configuration with different operating systems running the same (or even different) application server Test the system redundancy periodically Monitor traffic for abnormal behavior
Place all VoIP traffic into a different VLAN
When possible, enable TLS authentication and encryption for SIP signaling and use SRTP for media encryption

Thanks to its wide proliferation, Asterisk has become the target of many hackers. In its most simple variant, hackers scan networks looking for SIP hosts, then try to find valid extensions, and finally try to hack into the passwords of these extensions. It is therefore well worth discussing some elementary security steps that can help enhance its security posture. Some of the steps apply to Linux per se, while others are especially applicable to Asterisk itself. More details can be found in [5, 6].

- Enable and use IPtables. IPtables allows configuring the tables and rules for Linux kernel firewall. It can be thought as a software firewall, so basic security rules that apply to any firewall apply here too (e.g., blocking non-necessary ports). An even better solution is not to enable unnecessary services at all rather than having to block ports.
- Use a hardware-based firewall. Even a cheap hardware firewall can add a further layer of security to Asterisk, since they can filter the incoming malicious traffic from Internet.
- Use tools that protect against random password attacks by blocking IPs that make repeated unsuccessful password attempts.
- Use IP address restrictions. Not all IPs in your network have the same needs. Using the principle of least privilege, these IPs can have different access rules. For example, internal IPs used for telephony do not have a need to access the Internet directly. The other way around, external IPs (of remote sites, remote workers, etc.) can be explicitly permitted to enter the network while other IPs are denied. Of course, this implies that external users have static IPs.
- Change port numbers and use NAT port translation. For maintenance services such as HTTP, FTP, and SSH, it is always a good idea to change the default port numbers. By periodically changing these ports an extra layer of security can be created. Of course, this cannot be done for SIP and IAX ports; otherwise your PBX would not be able to talk to external PBXs and users.
- Disable web access to your PBX. Although the web graphical interface is a very user friendly way to configure the PBX, it is yet another entry point for fraudsters.
- Implement VPNs for PBX systems. VPNs are encrypted data tunnels that can offer secure communications in an otherwise insecure transmission path. The major challenge here is that both end points must be using the same VPN.
- Regularly check your logs (even daily). The more frequent the checks, the less time it takes for a compromised system to be revealed, minimizing thus the losses.
- Stay up to date. Regular reading and information gathering are needed to stay abreast of the security issues and vulnerabilities of your system.
- Disable "dialout" and "callback" options in voicemail. This way even if fraudsters get access to a mailbox they will not be able to make calls.
- Should you be using a "pay as you go" VoIP subscription, make sure you keep the available funds limited and do not enable the automatic credit card charge. This way, in case of a breach the loss will be limited to the available funds. By carefully replenishing the account you can always leave a "room" for extra communication costs, but should that extra budget gets used, then you are immediately alarmed that something is wrong. That is, you are alarmed that the communication cost has surpassed a limit and that could possibly be appointed to fraudulent use.

- If you have a mix of POTS and VoIP lines, don't put the POTS lines in the default outbound pool for toll calls. This way fraudsters are limited to using the VoIP lines and therefore the cost can be controlled as per the previous step.
- Disable international calls or if this is not feasible, add the PIN feature to any international call so that there is an extra "password" before making an expensive call.
- Set your Asterisk to reject bad authentication requests on valid usernames with the same rejection information as with invalid usernames. This way, attackers cannot get feedback on whether they are using a valid username.
- Block your AMI manager ports. AMI ports allow a client program to connect to an Asterisk instance and issue commands or read events over a TCP/IP stream. Make sure you limit the IPs that can connect to this port to only trusted ones.
- Limit the number of simultaneous-parallel calls per SIP entity. This way, in case of a breach, the number of fraudulent calls and along it the bill will be minimized.
- Limit the overall number of simultaneous-parallel calls. Same as the previous point, but concerning the whole PBX. For example if you have 100 users, it is realistic not to expect more than 40 calls to be concurrent (unless of course you are into tele-marketing or another heavy telephony usage market).
- Make your SIP usernames different than your extensions. A more technical idea is to un-map extension numbers form SIP identities. This makes it considerably harder for outsiders to find valid identities to attack via password guessing. It must be noted that such fine-tuning cannot be implemented using graphical interfaces.
- Ensure that non-authenticated callers cannot reach any toll trunks, or even cannot make any calls at all.
- A complex idea is to configure phones to "encode" the dialed number using leading and trailing suffixes and prefixes that are later dropped by the PBX itself. That way even if a hacker gets access, he will not be able to make a call since he will be missing the extra digits that the PBX is stripping.

4.9 Logs

Log maintenance and analysis are essential for the security of a system. The relevant data are written into the log files on the storage device, in ASCII format or in binary and compressed format. There are even PBXs that hold the logs in SQL-type databases for easier archiving and retrieving of information. In any case, most PBX systems provide embedded logging systems that can be customized in many different aspects. Although this is a good solution, an even better solution is to duplicate these logs, in real time, in another system. In embedded logging systems existing log files can be printed and erased by using file system commands. Using a standalone, separated call logging system provides good chances that a successful attack in the PBX itself, successful in deleting or amending logs, will not be able to do the same with logs that are hosted in a better secured and dedicated call logging system. It is also interesting to remember older technologies, where records were physically printed in real time, leaving a paper trail that was impossible to delete without physical access to the premises. Of course, such a solution is not practical nor feasible in a large PBX where thousands of records are created every hour.

Needless to say that call logging itself should be enabled. There is no point having a log server if it does not log any traffic! Having enough data from call detail records (CDRs), the administrators can use data mining techniques to spot fraud patterns and illogical behavior that could be attributed to malicious actions. An important step is to extract the typical call patterns of the organization. Hours of activity, most frequent destinations, most frequent numbers called, mean duration of calls, completion rate and retry behavior, etc. are only some of the information that can be associated with a given installation. This way, deviations can be spotted and immediately be researched for the possibility of abuse. As an example, if the typical call pattern is limited to business hours, then calls found late in the night or in the weekends are a warning sign. Other typical patterns that should be examined are very short incoming calls that point to war-dialing attempts. Very long outgoing calls (especially if coupled with equally long incoming calls via 800 numbers) should also be brought to attention since they could be fraudulent ones. Specifically for these types of call, some brands implement an automated alarming feature to alert the administrator about their presence.

Reconcilliating data from the in-house logging system with the data from the provider's logs is also handy. Inconsistencies should be checked as they could be attributed either to malfunctions, wrong billing, or even deletion of logs to cover criminal activity.

Apart from the CDRs, modern PBXs hold logs of logins, management, and practically they can be setup to log almost everything that takes place, as long as the necessary storing space is available. The amounts of data that need to be processed are growing in a fast pace. Huge data volumes are not easy to work with and to analyze. This is why software could help automate the process as much as possible.

Closing, a very important point that must be noted is that call logs get created after the end of the call. As long as the call is active, there is no respective record, since the record has to include the end-time and duration of the call. Such a call starting from a hacked PBX and staying for many hours online will not be presented in the CDRs should the administrator check during the time the call is active. It is therefore important to enable low level tools and logs that give insight to the status of trunks and their occupation load. This way, the administrator can see that a given trunk is unusually busy and further probe to see who is the user connected to it, where does the call goes to, and so on. We will see more details on logs in the forensics chapter that follows. Before that, Table 4.4 synopsizes this section.

4.10   The Most Important Tasks

It is quite impressive the fact that most PBX incidents do not involve hacking with high-tech hacking tools or sophisticated attacks. Hackers are using the same tricks that have been discovered 25 years ago, with the same tools and the same modus operandi. Yet, companies fail to protect their PBXs. Although simplistic in nature

Table 4.4 Important points about PBX logs

Enable logging for administration changes and for calls
Use a stand-alone separated call logging system Extract the organization's typical call pattern Check for deviations and fraud patterns Reconcilliate with provider's logs
Check trunks' occupation/load
Monitor central office lines utilization
Check administration logs for unauthorized changes and keep changes log
Use software to automate the process
Logs get created AFTER call ends
Enable monitoring and alert features for short or long holding times on trunks. Short are used in war dialing and password guessing. Long are used in compromised trunks

Table 4.5 The top 15 security measures list

Non technical
Enforce formal procedures and policies and educate users, administrators, and operators
Keep everything documented and tidy (equipment, cabling, connections, setup) Incorporate PBX into DRP and BCP plans
Keep contact with installer–provider–manufacturer
Consider implementing an information security management system such as ISO 27001
Technical
Use strong authentication (Password + {smart card, one-time tokens, callback, etc.}) and change passwords frequently
Disable remote management, use it only with callback and specific procedures
Disable not needed commands/options/features
Check settings, configuration and administrator and user privileges Restrict international and premium rate numbers calling capability Enable logging and check call and administration logs
Disable DISA or restrict to local calls only
Pay attention to voice mail and the automated attendant
Consider a PBX firewall
Consider the use of hardware cryptophones or software suites to encrypt communication

the measures described can actually minimize the threat PBX face and should not be treated as "trivial" ones. Once again, user awareness is among the most critical imperatives and hopefully this book achieves this goal.

To sum up everything we have discussed in this chapter, a list with the top 15 security measures that should immediately be taken is presented in Table 4.5. The interested reader can find more details in [7, 8].

4.11 Advice for Simple Users

Although administrators implementing the measures we described can handle most of the problems, simple users have also a great share in assuring the PBX security. We present a few points for them to keep in mind in Table 4.6. Interested administrators could organize a short educational seminar to inform their users about these issues.

Table 4.6 Security advices for PBX users

Do not leave your phone-set unlocked
Keep your all service security codes (PINs, etc.) secret and change them frequently
Write down the serial number of your phone-set
Personalize your phone-set so it can't be swapped with another one
Pay attention in phones present in meeting rooms and during confidential discussions
Do not save sensitive data, e.g., PINs or other codes or mobile phones in the memory of your phone
Familiarize yourself with the screen's indicators-icons-tones of your phone and pay attention to them (e.g., intrusion tone)
Familiarize yourself with call waiting and conference call announcement tones (beeps heard periodically during the call)
Report repeated wrong dialings and silent hung ups
Do not blindly trust the caller ID or the originator of a call as it might be spoofed

Closing this chapter, in the last section we will present the idea of a collaborative project, aiming at protecting PBXs in a nationwide scope.

4.12 On a Collaborative Project: PRETTY (PRivatE Telephony SecuriTY)

Like other initiatives, such as ENISA Expert Group on research priorities in the areas of networking and information security for resilient networks [9], a renewed focus on PBXs with a coordinated project can guide PBX networks to becoming more resilient. Along these lines, a targeted, centralized action in order to both educate the users and secure their telephony systems is proposed in "PRETTY" [10]. It compromises of educational, policy, auditing, technical, documentation, hardware and software solutions and actions that could be implemented in a transnational project. The project's framework consists of four main work packages, as described below. Interested readers are most welcome to join the author into actually implementing this project idea.

4.12.1 User and System Requirements

In the early phase of the project, the requirements will be set up focusing on confidentiality, integrity, and availability of PBXs as well as ease of use for the operators. As a first step, the requirements of PBX systems from an operational point of view will be collected and analyzed. Furthermore, an un-elastic requirement will be not to hinder at all the PBX's functionality after the implementation of the software and hardware tools proposed.

4.12.2   Research and Development

A substantial part of the project will be devoted to research and development of methodologies, software and hardware. A model consisting of rules and situation descriptions will be researched in order to develop an expert system that will assist in creating and maintaining higher security for the operation of a PBX system through adoptable guidelines for PBX operators. It will encompass standard operating procedures, security policies, disaster recovery plans, and methodologies to help dealing with fraud, detection, and prediction of maintenance problems as well as auditing.

The different aspects of securing the telecommunication providers services will also be studied. Specifying the possible attacks and fraud scenarios and approaches will help preventing and handling those misuses. Detection and prevention techniques will be developed, leading to appropriate mechanisms and a framework for providing secure infrastructures. Furthermore, applicability of state-of-the-art security solutions will be evaluated for their usability for telecommunication providers' services protection.

Finally, apart from software solutions, a cost-effective hardware PBX firewall will be designed in order to provide an extra layer of security, irrespectively of operators' behavior. In any case, it should be noted that the methodology used will be able to cover both classical PBXs and VoIP ones.

4.12.3   Implementation

The implementation phase will lead to prototype installations and solutions, stemming from the results of research. In order to test the efficiency of these means and recommendations, the software and hardware solutions will be implemented in a selected number of PBXs serving banks, hospitals, public bodies, ministries, industries, universities, etc. In addition, the expert system and guidelines will be presented to these PBXs' operators. The security level increase will be demonstrated by the security audits performed before and after implementation. This phase will cover among other, all the security aspects presented in the previous chapters of the book.

4.12.4   Dissemination of Results

An equally important part of the work, the dissemination of results will aim at the broadest possible audience. The expert system will be publicized and educational material will be distributed to ministries of telecommunications and to telecom providers. Actions to raise awareness will include conferences and seminars, targeted for both specialists and the public.

4.13 Conclusion

Much has been said and done regarding data communication security but PBXs have been left unprotected, forgotten, and waiting to be attacked. Checking the proper operation and ensuring the safety of PBX as well as protection against unauthorized use and access are usually left to the owner. This has also been stated in various security manuals of vendors. This has of course tremendous effects since due to economic and technical difficulties, in essence it is impossible to guarantee that the proper measures are taken. This chapter outlined educational, policy, auditing, technical, documentation, hardware and software solutions and actions, in order to both educate the users and secure their telephony systems. Along with a proposal for a centralized and focused project, it will hopefully form the basis for actual implementations by various stakeholders.

References

1. deDECTed. https://dedected.org/trac
2. IT Governance, ISO 27001 & Information Security. http://www.itgovernance.co.uk/iso27001.aspx
3. SecureLogix Corporation (2007) White paper: voice network management best practices, March 2007
4. Prevelakis V (2007) The Athens affair. IEEE Spectrum, July 2007
5. Mundy W. Avoiding the $100,000 phone bill: a primer on asterisk security. http://nerdvittles.com/?p=580
6. Todd J. Seven steps to better SIP security with asterisk. http://blogs.digium.com/2009/03/28/sip-security/
7. Archer K, White GB et al (2001) Voice and data security. Sams Publishing, Indianapolis, IA
8. NIST (2001) PBX vulnerability analysis. Special publication 800-24
9. Rauscher K (2009) ENISA Expert Group on research priorities in the areas of networking and information security for resilient networks. ENISA, Athens, Greece
10. Androulidakis I (2012) PRETTY (PRivatE Telephony securiTY)—securing the private telephony infrastructure. Inf Secur Int J 28(1)

Chapter 5
PBX Forensics

5.1 Introduction

The right use of evidence forms an important judicial process that significantly helps a case's hearing. The digital forensics analyst assembles evidence from the crime scene, evaluates its importance, and analyzes and presents the data in the court. In analogy to "classical" forensics, digital evidence analysis takes place using data extracted from any kind of digital electronic device [1]. On one hand, PBXs can be a "victim" in a fraud case. Therefore, the analyst has to extract as much data as possible in order to locate the fraudsters. On the other hand, in case a person is involved in an illegal activity and has used a PBX, it is more than likely to have left digital traces which constitute valuable material for the law authorities. As time goes by, more and more data are collected from their use, strongly correlated to their owner. Such data can be found in the PBXs themselves as well as in the computer systems of the telecom providers' networks and infrastructure.

5.2 Crime and PBXs

The total number of PBXs (TDM and VoIP) only grows. Obviously, criminals are keen to target such a great platform. Fraudsters are using the technology for personal gain and "involve" PBXs in various ways in illegal and criminal activities. Apart from direct financial gain, PBXs can be used as a means of anonymous communication. Indeed, with the banning of anonymous mobile phone service with prepaid cards, PBXs will again become a very attractive target for the exchange of information among criminal rings. PBX boards, with their small footprint and increased monetary value, are a promising target for thieves.

At the same time, focusing on "white collar" crime level and with the PBXs in presence, we observe all kinds of telecommunications fraud, personal data theft,

© Springer International Publishing Switzerland 2016
I.I. Androulidakis, *VoIP and PBX Security and Forensics*, SpringerBriefs
in Electrical and Computer Engineering, DOI 10.1007/978-3-319-29721-7_5

identity infringement theft, eavesdropping and industrial espionage (using the phone itself as a bugging device to intercept voice and commercial secrets), and so on. We provided some data on telecom fraud in the first chapter, with losses of many billion dollars per year. The modern commerce options using telephones to perform banking transactions and services will obviously make the problem more immense, providing more space for criminal activities.

5.3 The Warning Signs

Judging from the previous, it is a safe assumption that any given PBX will sooner or later be targeted by an attacker with a crime in mind. Before getting an enormous bill, where it is already late, there are some signs that when manifest, administrators should immediately proceed to a forensics analysis. The same analysis should take place if an incident is already confirmed.

One of the first signs of telecom fraud is incoming or outgoing trunks overload. Trunks are permanently busy, users complaining they cannot get a line to the public network, and clients complaining they cannot reach the company. In case the company offers toll free (e.g. 800 numbers), an increase in its load is also a sign that should be examined. Indeed, an 800 line is the most sought after for abusing, since the fraudster will avoid even the local call cost.

A change in the call patterns is also a hint, with increase in international calls, in new destinations that the company normally does not have business with. Lengthy calls or calls to premium rate services during late night hours or weekends and holidays are typically found in a breached PBX. Being able to excerpt the typical call pattern shows the importance of call analysis that will be soon analyzed.

Before the actual exploit, there might be a surge in silent short calls, or calls from persons claiming they have got the wrong number. Both are signs of war dialing taking place, either manually or automatically. In addition to that calls from persons claiming they are with the provider or with the telecom company are a sign of social engineering attacks.

Specifically for the voice mail, strange messages left in boxes should be examined. Obviously, new mail boxes instances that were not activated by the administrator, as well as presence of calls initiating from the voice mail system are signs that something is already wrong.

Closing, in the confidentiality front, leakage of business secrets and intellectual property could be a sign that phone conversations have been intercepted.

5.4 The Hacker's Modus Operandi

A PBX can be attacked in a number of ways including: physical access, abusing remote maintenance–management or remote access functionality, features misuse, and services misuse (DISA, IVR, ACD, AA, voice mail).

We have seen in Sect. 3.16 a malware attacking PBXs. Along the same line, a hacker would work with the modus operandi we will describe in this section. It is important to familiarize with the way fraudsters work in order to be able to better secure the PBX and follow the forensics trails after an incident. As such, based on the flowchart of Fig. 5.1 the hacker would do the following.

1. Pick up the target (either a specific one or a random one).
2. Do a thorough search in the yellow pages and in the Internet for documented lines (directory of phones, direct dial in lines, etc.).
3. Proceed to war dialing, dialing all of the numbers in the specific numbering plan.
4. This step is usually performed with automated tools but can be accomplished with manual dialing too. A simple spreadsheet sheet as in Fig. 5.2 can be used in order to note down the details.
5. Judging by the tone and the pattern of the ring tone it might be possible to determine the type of the PBX. Most of the times, the music on hold theme is a clear indication of the PBX manufacturer.
6. When a modem is found then its prompt can help evaluate the type of the equipment connected. It might be a server, the PBX maintenance port, or an employee's PC. The hacker will proceed relevantly.
7. Should the PBX maintenance modem is found, then the default passwords would be the first ones to try. Otherwise guessing can have some success, while social engineering would probably also work. The login screen is usually identifying the concerned system.
8. Once inside the system, the attacker will initially deactivate logging features. He will then proceed to probing the system in order to understand its configuration. It is then relatively easy to create virtual numbers, to activate features such as DISA or immediate forward to premium rate service numbers, to download call records and do whatever he pleases. Furthermore, systems tend to be user friendly which is not always a good idea. In our case even if he does not know the exact arguments for the needed commands, an online help system invoked by typing "help" or special characters such as "?" could allow him to complete his actions.
9. If a service is found (DISA, voice mail, IVR) then it can also reveal information about the type of the PBX. Furthermore, each type has documented features and problems that a malicious hacker might try to exploit.
10. Daring enough hackers could also physically present themselves to the PBX site, posing as technicians and asking to visit the PBX itself in order to proceed to "maintenance" works.

The previous steps can be assisted by social engineering tricks where the attacker will manipulate the human element in order to divulge valuable information as already mentioned before. Social engineering techniques, convincing users to do something while the attacker is possibly masquerading as somebody else (colleague, boss, technician, etc.), play an important role in the success of these attacks.

Alongside the previous modus operandi, there is also the possibility to engage into a denial of service attack. Recapping these attacks, apart from obvious methods such as shutting down the PBX or stealing vital parts of it (such as the CPU), there

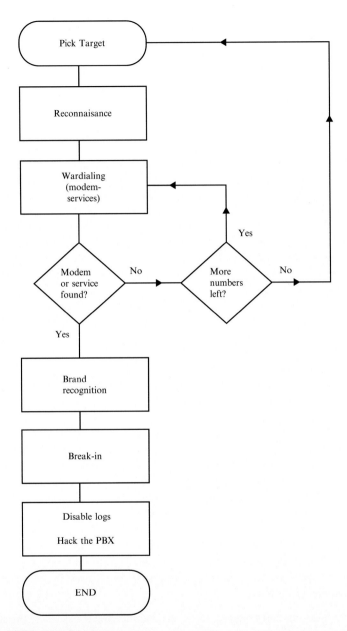

Fig. 5.1 PBX hacking flowchart

are also less apparent techniques. One would be to instruct an array of different PBXs to start calling the numbers of the victim PBX. It would overwhelm it with so many calls that legitimate users would not be able to access it. Inserting deliberate faults (e.g., in the routing tables) and degradation of the PBXs capabilities (probably by reducing the number of available trunk lines) can also cause significant problems to the operation of the targeted company or organization.

ABCD	0	1	2	3	4	5	6	7	8	9
00										
01										
02										
03										
04										
05										
06										
07										
08										
09										
10										
11										
12										
13										
14										
15										
16										
17										
...										
...										
...										
...										
...										
...										
...										
80										
81										
82										
83										
84										
85										
86										
87										
88										
89										
90										
91										
92										
93										
94										
95										
96										
97										
98										
99										

Fig. 5.2 Manual war dialing matrix

Closing, it is interesting to note some extra details in regard to yellow pages (point 2 in the previous page) taken from non-published research from the author. The Greek yellow pages for capital Athens, back in 1996, contained 1,817,075 telephone entries. Table 5.1 presents an interesting breakdown of these numbers, where it is immediately apparent that PBX can easily be targeted. Quite interestingly, even modems were listed in some occurrences!

5.5   The Evidence

As stated, PBXs provide a continuous flow of data and information concerning their users and their behavior. Where all this information resides to? Not only in the PBX itself but also in the computer systems and networks of the providers–carriers.

Table 5.1 Athens yellow pages (1996)

1,817,075 numbers in total
• 2319 listed as PBXs (head number)
• 248 PBXs with 10 or more lines
• 133 PBXs listed as DDI (direct dial-in)   – These are the "big" ones with more than 100 lines usually
• 22,069 belonging to ministries
• 14,949 to banks
• 7709 to industries
• 2642 to hospitals
• 1851 to national PTT
• 819 to public power corporation
• 739 to Public Water Supply and Sewerage Company

Specifically for the PBX, evidence can be found in its internal memory and the hard disk drive (HDD). We should not forget the case where PBXs are maintained using a computer connected to them; evidence can be found in the management computer too.

Typical data to look for than can prove suitable evidence are the following:

- Phone catalogues and contacts
- Incoming/outgoing/missed calls
- Incoming and outgoing text messages (many PBXs allow the exchange of short messages like mobile phones do)
- Entries in the voice mail (messages left, sound recordings, voice notes)
- Alarm clocks/reminders, wake up calls
- User identifiers (e.g., phone lock codes)
- Device identifiers (e.g., serial number of the phone)
- Catalogue of providers–carriers used
- PBX services used
- Geographic information (e.g., destination of calls)

A particularly important differentiating factor concerning computer forensics is the fact that PBXs were traditionally more "closed" systems, allowing less user control to their core functionality and their operating system. It was thus difficult to precisely know what data were stored where. Things have changed with open source VoIP solutions.

5.6   Fundamental Questions and Problems

As it is the case with traditional evidence and forensics, the specialized professionals should follow strict and specific techniques for data extraction, maintenance, analysis, and presentation of digital evidence. Some fundamental questions arise during the process.

The main question to answer is whether the given telephone was used for criminal/ illegal activity. In case the answer is positive, can we extract relevant data to be used as evidence? The other way around, are there data pointing to a fraudster that has abused a PBX?

Certain problems manifest in the whole process: was the specific suspect the phone user? In other words, the telephone number of the owner was involved in the crime… but who was the user at that moment? Was it its owner or a third person? What happens if the number was spoofed? How can it be proven that the phone was not used by another person? In addition, how is it proven that the incriminating evidence stems from the phone owner's use and not from some other malicious activity (a virus or malware)? As we have seen, call selling operations are a serious business. An interesting call traced could unfortunately point not to the fraudster, but rather to the person who bought the service, most possibly unbeknown to him that was taking part in a fraud.

It is also possible that records in telecom operator's systems are not complete, or that have not been logged at all (because of system overload, malfunction and so on). Maybe the data retention period for preservation of data has elapsed resulting in evidence being lost. Finally what is the possibility of a simple human error (e.g., a wrong call at the wrong time to the wrong person, incriminating a completely innocent person)? Even though it may resemble to a film script, statistically everything is possible.

Apart from the digital space, the examination should take into consideration the preservation requirements of other (not digital) evidence which can coexist (DNA, imprints, drugs, guns, etc.). Therefore, the examination sequence is important (e.g. touching a device with naked hands can destroy fingerprints and reversely, in the process of extracting fingerprints from a device, it is possible to accidently delete digital evidence that potentially existed (e.g., by accidently pressing the redial button).

The reader can now understand how crucial the evidence analysis procedure is and how careful the analyst must be. She has to examine every possible parameter that led to the presence of that specific evidence in the device. Simply extracting data is not enough. Only a multilayered and thorough correlation combining data from various sources (e.g., payphone logs, prepaid card logs, credit card logs, etc.) leads to usable evidence.

5.7 Forensic Procedures

5.7.1 Introduction

Having stated the framework of forensics, in this section we will be focusing on the nontechnical details of digital evidence collection and analysis. We will cover procedures for evidence collection, preservation, examination, and analysis and finally for the writing of the report. We will rely on the best practices that IOCE (International Organization on computer evidence guidelines for best practice in the forensic examination of digital technology) proposes for digital evidence. They state the following [2]:

- When dealing with digital evidence, all of the general forensic and procedural principles must be applied.
- Upon seizing digital evidence, actions taken should not change that evidence.
- When it is necessary for a person to access original digital evidence, that person should be trained for the purpose.
- All activity relating to the seizure, access, storage, or transfer of digital evidence must be fully documented, preserved, and available for review.
- An individual is responsible for all actions taken with respect to digital evidence while the digital evidence is in their possession.
- Any agency which is responsible for seizing, accessing, storing, or transferring digital evidence is responsible for compliance with these principles.

We will also use Association of Chief Police Officers (ACPOs) guidelines which in turn has also published its own best practices on the object (good practice guide for computer-based electronic evidence) focusing on four principles [3].

1. No action taken by law enforcement agencies or their agents should change data held on a computer or storage media which may subsequently be relied upon in court.
2. In exceptional circumstances, where a person finds it necessary to access original data held on a computer or on storage media, that person must be competent to do so and be able to give evidence explaining the relevance and the implications of their actions.
3. An audit trail or other record of all processes applied to computer-based electronic evidence should be created and preserved. An independent third party should be able to examine those processes and achieve the same result.
4. The person in charge of the investigation (the case officer) has overall responsibility for ensuring that the law and these principles are adhered to.

5.7.2   In General

During PBXs' forensics examination, the evidence should always be extracted maintaining the data integrity. General principles of forensics that concern classic evidence are also in effect for digital evidence. Reliable procedures for securing the scene, documenting–photographing it, evidence collection and proper packing, transfer, and storage should be followed. Each action relative to the data confiscation, access, transfer, and storage should be completely logged and available for possible audit. During the confiscation, the area and evidence are photographed and/or videotaped.

After confiscation, the evidence should not be altered. Consequently, the analyst should have proper training while the procedures she follows should be possible to be repeated producing the same results in order to prove their validity. At the same time, in all stages, each evidence holder is responsible for anything that happens to it. Finally, the procedure should be completed as quickly as possible because if the evidence remains for a long period in the forensics laboratory, there is the danger of alteration accusations from the defended.

5.7.3 Training and Competence

All involved persons in the process should be suitably trained in order to perform and document their actions. Modern technology evolves so fast that using the best equipment alone does not guarantee the success if it is not accompanied by regular and thorough education of the examiner.

There are different kinds of responsibilities and training, according to the involved personnel. Police officers who act on the scene of the crime have different responsibilities than analysts. The first ones should secure the scene, log and collect the physical evidence (e.g., telephone sets), pack and safely transfer them to the laboratory. They shouldn't do the analysis themselves since they do not have the necessary know-how and the specialized equipment. The analysts, who are the ones responsible for the analysis, perform the data extraction and the examination of it, finally writing the corresponding report. They are obligated to remain informed and to follow the technology pace and the scientific research evolution. In any case, their work doesn't stop in the report but they must also safeguard the evidence for possible future need and access.

The complete responsibility for the suitable procedures observation lies with the Head Officer. He must ensure that all the involved personnel have the proper training in order to complete the corresponding procedures. Following, during the analysis, he must maintain continuous communication with the analyst and provide guidance since the Head is the only one who knows the whole case. If this communication is disturbed it is possible to lose data which can have decisive importance for the trial. On the other hand, the expert in charge of the forensics research decides for the suitable methodology she will follow depending on the case's severity.

5.7.4 The Analysis Procedure Itself

During the procedure of data extraction different situations are possible. PBX access might be impossible (therefore the only data that can be used are based on the provider's help), can be temporary, or in the best case scenario the PBX and phones have already been confiscated and are in the examiner's laboratory (or at least the scene is secured and the analyst can work in the actual infrastructure). A major differentiation here, when dealing with PBXs, is that due to their big volume and size, in most cases it is practically impossible to confiscate them. This is why a better option is to seal the place and wait for the analysts to come on site. Analysts can perform a digital, bit for bit dump, cloning the PBX's HDD and work with the copy. Although practical, this process leaves behind evidence stored in memory chips of the different PBX boards. Hopefully data in the HDD will be enough to solve the case. There are also flash memory cards that should be cloned too. On the other hand, there are PBXs that do not have HDDs. Luckily, this probably means that they are low traffic ones, serving few users. As such, they are small in size and it might be after all possible to physically transfer them to the laboratory. The interesting point here is that they actually can be transferred while still on power, using either their included batteries or an external UPS.

The PBX is initially recognized and the selection of tools and methodologies that will be used follows. The expert seeks information for whom, where, when, and why by examining data stemming from the PBX sets ownership but also from their possession (it is possible that the phone has a different owner than the user of the phone at the moment of the confiscation). This is performed by examining data, applications, and files, categorizing the time line of events, without overlooking hidden evidence that possibly exists.

As a general principle, it is good to photograph or videotape each step of the process, time stamping it. The examiner's actions are recorded in the corresponding log. The log includes details for the time, the action, and its result. A third person following the same instructions should reach to the same result. Even for automated data analysis (when the PBX is based on a known/open and not closed/proprietary O/S), videotaping is essential in order to prove that the correct way of operation was used. It can also be used to verify the results. The tools and the version of software used should also be documented and included in the final report.

5.7.5 Data Preservation and Isolation from the Network

The main principle of forensics is the preservation of data, indicating that data should not be altered at all to allow its use in Court. For this purpose, it is advised to solely use special software as non-qualified and unchecked software can potentially write/alter data destroying the integrity of the evidence. Contrary to the IT domain, very few specialized tools exist to aid in a PBX forensics procedure. In VoIP systems, things are better since classical IT forensics tools can be used. In any case, a major (but not the only one) part of the forensics analysis deals with logs analysis, where a multitude of products can be used. Leaving logs aside, PBX may often not be compatible with specialized software or even generic software. In such cases operating methods and tools should first be checked on a same model test PBX, a difficult and expensive process. In addition, the non-specialized software should be used as late as possible in the process.

A case can literally be "lost" due to mistakes in receiving and transferring data that destroy evidence. At this point, the examiner has to make an important decision. Should he leave the PBX working, as to monitor and possibly trace the fraudsters or should he isolate it from the network to prevent any data contamination? In most cases, dealing with ongoing economic crime and fraud, or with ring operations that use the PBX as a covert communication means, a secret operation silently monitoring the PBX can lead to the criminals. This requires the PBX to keep operating, as a "honeypot." In cases, however, that the PBX was used in a one-time crime, such as a murder (either the killer or the victim being users of a PBX), then it is particularly important to avoid "contamination" from new phone calls, messages, and generally communication with the network. New data may arrive (e.g., calls or messages) that can fill buffers and overwrite precious evidence. For this reason, the PBX should be isolated from the network. Unplugging the trunk circuits is not such a good solution since error messages will fill the respective logs. Circuits should be taken down either using the proper PBX commands or even better in collaboration with the operator. This way, the operator can isolate the PBX from the network, while the lines still appear to work from the PBX's side. As such no errors will be logged but at the same time the PBX will remain intact.

5.7.6   Identification of the PBX

Once the PBX has safely arrived in the forensics laboratory (or, once the analyst has arrived in the scene to work locally), the first task will be to identify the exact model and type in order to assemble the necessary connection cables, manuals, and so on. The majority of the market is dominated by well-known brands that cater for most installations. Given that, a simple glimpse is often enough for the identification of the PBX. It is possible, however, that the model is unknown or does not exist in the local market.

After having identified the phone the examiner searches for its user manuals and its technical characteristics. Having read the manual (actually, having read dozens of manuals of different PBXs in her career) and being familiarized with the environment, the examiner is in place to proceed to the next steps. She will define the storage area and extraction capabilities and will locate the available control and data interfaces. Adhering to the security and data integrity requirements, the examiner has to understand the device design and explore both electronic and mechanical parts and interfaces. At the same time, she will calculate the time needed and the cost of the procedure before finally extracting the data.

5.7.7   Examining the Evidence

The actual task of the examination of evidence contains the analysis of call logs, user logs, changes logs, phone sets' and PBX's memory contents, trunk connections, signaling exchanged, etc. We will see more details in the following sections. The work takes place on a copy of the original exhibit (a clone of the HDD usually), or in the PBX itself (for smaller PBXs that do not have a HDD). For the HDD clone the examiner calculates the corresponding hash. The hash is a unique fingerprint-signature for every file, achieved by special algorithms. Checking the hash values of two different files or data sets, we can easily certify that they are the same and that there are no differences at all, not even in a single bit.

Following the extraction of evidence with the use of automated tools, a manual effort can verify the results and sometimes further discover evidence that for various reasons (e.g., insufficient device support) the software failed to find. Photographing or filming the steps is highly recommended again. Needles to say that in cases where no compatible software for the analysis can be found, the manual analysis method may be the only option available.

In conclusion, the evidence extracted must be verified. If possible, a cross-check could be sought with the network operator. For example, calls registered in the call logs can be verified against the calling logs of the network provider–carrier. Using scientifically proven procedures and thorough analysis, the examiner increases the importance and weight of the evidence since it can no longer be disputed.

5.7.8 Findings Report

The process ends with the report of findings which records both the process followed and the findings themselves (actually the process does not end here, since evidence has also to be preserved for as long as needed, but this is out of the scope of the book). In this report initial data for the Agency, the case, the staff who dealt with it, and the Head as well as the dates of the events should appear. The presentation of the software and tools used and the methodology followed comes next. Moreover, the planning of the methodology and the steps taken must be clearly formulated. The accompanying materials and equipment used are recorded in the relevant section. They are presented after the findings from both the automated extraction and the manual investigation, and the report closes with conclusions. A good report will assist in the proper trial so the analyst should possess advanced writing skills too. In fact, an excellent technical part of the forensics process may lose its value if not presented in a well-intelligible way on the report.

At this point, we will switch to a technical level, proceeding with details about logs, real-time data, and evidence stored outside the PBX.

5.8 Logs

5.8.1 In General

Back in Chap. 4 we discussed about logs. In this section, we will see some more technical examples of logs that are closely related to the forensics process. Namely, we will examine the command log, the authentication and system log, the alarms log, and the calls log.

5.8.2 Commands Log

The commands log enables the analysis of terminal activities. It can also be edited and used to regenerate the data that were changed since the last saved log, or to synchronize data between the system and an external data base. A command log

```
----------- Log file started <19/03/12 10:14> -----------
SET Subscriber    "101": "7167"
{
 Public_Network_Category_Id = "30"
}

----------- Log file started <19/03/12 10:15> -----------
SET Subscriber    "101": "9157"
{
 Public_Network_Category_Id = "1",
 Display_Name = "TEST"
}
<101>xa00101> _
```

Fig. 5.3 Management log example

```
Mar 12 12:26:15 xa00101 HISTORY[18133]:  ppid=18132 UID=2011 cmd=mgr
Mar 12 13:11:59 xa00101 HISTORY[18016]:  ppid=18015 UID=2011 cmd=exit
Mar 12 13:23:18 xa00101 HISTORY[20650]:  ppid=20648 UID=2011
Mar 12 19:35:10 xa00101 HISTORY[24393]:  ppid=24392 UID=2011 cmd=ps
Mar 12 19:35:13 xa00101 HISTORY[24393]:  ppid=24392 UID=2011 cmd=ps -ef | grep sh
Mar 12 19:35:14 xa00101 HISTORY[24393]:  ppid=24392 UID=2011 cmd=exit
Mar 13 11:23:24 xa00101 HISTORY[1843]:  ppid=1842 UID=2011 cmd=mgr
Mar 14 11:13:30 xa00101 HISTORY[16529]:  ppid=16528 UID=2011 cmd=readkeys
Mar 14 11:13:34 xa00101 HISTORY[16529]:  ppid=16528 UID=2011 cmd=
Mar 14 11:13:43 xa00101 HISTORY[16529]:  ppid=16528 UID=2011 cmd=listtool | grep keys
Mar 14 11:13:51 xa00101 HISTORY[16529]:  ppid=16528 UID=2011 cmd=readkey
Mar 14 11:15:07 xa00101 HISTORY[16529]:  ppid=16528 UID=2011 cmd=mgr
```

Fig. 5.4 Commands log example

stores records of the entered commands, their execution result, and related data. It is also possible to log a periodic system dump. Command logging can be enabled as easy as issuing a simple command (e.g., IOELI:PRINT = YES, RESULT = YES; in a given brand). Figure 5.3 shows a typical example of such a log, from another brand. The user can note that logs are time stamped. In the first case, extension with number 7167 had a change in its public network category ID (i.e., the calling capabilities). One minute later, extension 9157 had a similar change, to another category ID, plus the displayed name in the internal caller ID of the PBX was changed to "TEST." In case an incident has taken place, the administrator can locate the changes that caused the incident. She can, e.g., find that a specific set was forwarded to an external destination.

Not only management changes but also all the commands issued in the man–machine interface or the O/S of the PBX can be logged. Figure 5.4 shows a relevant log. Again, all the commands issued are logged and time stamped.

A third type of log that can help in case a fraudster has managed to delete these logs is the log that is used to communicate changes in a network of PBXs. In such networks, a change in a given node must be communicated to other nodes. A new user in PBX A should be made known to PBX B, otherwise users of PBX B would not be able to call her. Each manufacturer has its own way of managing this interexchange of data among the PBXs. In many cases, a log is maintained that logs the changes of PBX A that are to be communicated to PBX B. Unfortunately, such logs hold far less information than local logs (not every local change needs to be communicated to the network) and are usually kept for much less time, nonetheless the attacker might have overlooked to delete them.

```
Mar 19 10:23:10 xa00101 xinetd[648]: START: telnet pid=24788 from=█████████.24
Mar 19 10:23:12 xa00101 PAM_unix[24789]: (system-auth) session opened for user ██ by (uid=0)
Mar 19 10:23:12 xa00101 ──██[24789]: LOGIN ON pts/0 BY ████ FROM ████████.24
Mar 19 10:25:13 xa00101 PAM_unix[24936]: authentication failure; ████(uid=████) -> root for system-auth.noradius service
Mar 19 10:25:23 xa00101 PAM_unix[24937]: (system-auth.noradius) session opened for user root by ██(uid=2011)
Mar 19 10:25:51 xa00101 PAM_unix[24937]: (system-auth.noradius) session closed for user root
```

Fig. 5.5 Authentication log example

```
Mar 19 10:00:01 xa00101 crond[24260]: (root) CMD ( /sbin/rnmod -as)
Mar 19 10:00:01 xa00101 crond[24263]: ████ CMD (/usr/sbin/logrotate -████████████████████)
nf)
Mar 19 10:01:01 xa00101 crond[24274]: (root) CMD (run-parts /etc/cron.hourly)
Mar 19 10:03:07 xa00101 PAM_unix[24320]: (system-auth) session opened for user ███ by (uid=0)
Mar 19 10:03:07 xa00101 ──██[24320]: LOGIN ON pts/0 BY ███ FROM ████████████.24
Mar 19 10:09:49 xa00101 PAM_unix[24320]: (system-auth) session closed for user ███
Mar 19 10:10:01 xa00101 crond[24497]: (root) CMD ( /sbin/rnmod -as)
Mar 19 10:10:01 xa00101 crond[24498]: ████ CMD (/usr/sbin/logrotate -████████████████████)
nf)
Mar 19 10:12:05 xa00101 PAM_unix[24514]: (system-auth) session opened for user ███ by (uid=0)
Mar 19 10:12:05 xa00101 ──██[24514]: LOGIN ON pts/0 BY ███ FROM ████████.24
Mar 19 10:20:01 xa00101 crond[24743]: (root) CMD ( /sbin/rnmod -as)
Mar 19 10:20:01 xa00101 crond[24744]: ████ CMD (/usr/sbin/logrotate -s /DHS3dyn/incid/logrotate
nf)
Mar 19 10:22:23 xa00101 PAM_unix[24514]: (system-auth) session closed for user ███
Mar 19 10:23:12 xa00101 PAM_unix[24789]: (system-auth) session opened for user ███ by (uid=0)
Mar 19 10:23:12 xa00101 ──██[24789]: LOGIN ON pts/0 BY ███ FROM ████████.24
Mar 19 10:25:13 xa00101 PAM_unix[24936]: authentication failure; mtcl(uid=2011) -> root for syste
Mar 19 10:25:23 xa00101 PAM_unix[24937]: (system-auth.noradius) session opened for user root by █
Mar 19 10:25:51 xa00101 PAM_unix[24937]: (system-auth.noradius) session closed for user root
```

Fig. 5.6 System log

5.8.3 Authentication–Logon Log

The authentication–logon log enables the analysis of system access. A logon log
stores records of logon events, both successful and unsuccessful, and logoff events.
In some systems logging of logon/logoff events is automatically initiated; it does
not have to be initiated via a command. For each logon event (or logoff) the following data are most often logged: The date and time of the event, the identification of
the port that was used to logon, the name of the user account that was used to logon,
the authority-privileges level associated with the account, the logon indicator (entry-
login or exit-logout), the logon result (successful or not), and the remote IP address
(connected via IP). Figure 5.5 shows a relevant example. This log can point towards
the way the intruder entered the system. If a local console session is registered, then
the attack started from the premises of the PBXs. Otherwise, an IP address can point
to the previous system that was used to connect to the PBX. In case of a dial-up
entry, the call detail record (CDR) will point to the telephone number used. Tracing
back IPs and telephone numbers most probably will involve the telecom operator.
Due to lack of space we will not extend with the description of this process. What
should be noted, however, is that most times, the IP or the telephone number will not
belong to the attacker, but rather to another previously compromised system. It is
well known that attackers use as many and as different as possible PBXs and compromised systems to hide their trails.

   In any case, the authentication log is usually part of the system log. In UNIX–
Linux like O/Ss, any important information regarding the operation of the system is
logged, in the system log as shown in Fig. 5.6.

```
20/02/12 22:15:46 001001M!001/14/0/007!=5:2053=Terminal 7 in service
22/02/12 06:54:34 001001M!001/04/0/017!=3:2050=UA Terminal 17 Loss
22/02/12 08:08:34 001001M!001/04/0/017!=3:2050=UA Terminal 17 Loss
22/02/12 08:08:42 001001M!001/04/0/017!=5:2053=Terminal 17 in service
22/02/12 08:09:12 001001M!001/04/0/017!=5:2053x002=Terminal 17 in service
23/02/12 12:04:51 001001M!004/03/0/019!=3:2050=UA Terminal 19 Loss
23/02/12 12:12:13 001001M!004/03/0/019!=5:2053=Terminal 19 in service
23/02/12 12:12:40 001001M!004/03/0/019!=3:2050=UA Terminal 19 Loss
23/02/12 12:17:08 001001M!004/03/0/019!=5:2053=Terminal 19 in service
24/02/12 08:43:42 001001M!001/04/0/023!=5:2053=Terminal 23 in service
24/02/12 09:15:40 001001M!002/05/0/031!=5:2053=Terminal 31 in service
24/02/12 18:33:27 001001M!003/02/0/030!=3:2050=UA Terminal 30 Loss
24/02/12 18:33:37 001001M!003/02/0/030!=5:2053=Terminal 30 in service
24/02/12 20:39:46 001001M!001/22/0/001!=5:2053=Terminal 1 in service
<101>xa00101>
```

Fig. 5.7 Alarms–incidents log filtered to show terminal losses and reconnections

5.8.4   Alarms Log

Apart from the system log that lists all critical operation for the O/S, there is usually a log to hold critical situations and incidents for the PBX itself and its functionality. Most PBXs have such alarm logs where malfunctions and other critical messages get logged. In Fig. 5.7, we have a filtered output, listing the exact times where sets have been set in service or out of service (this could be the case where the user unplugs the phone, or when a phone is stolen).

5.8.5   Calls Log

Proceeding, calls log (consisting of CDRs) is the most valuable evidence since it contains a wealth of information. As seen in Fig. 5.8, a typical record can contain almost 50 attributes for a given call.

The record contains the calling and called numbers, the date and time of the start, and the end of the call along the duration too. The duration can further be analyzed to call waiting duration and actual communication duration. This way, the administrator can have a clue about how much time did the phone ring before it was actually answered. The name of the user of the internal phone as stored in the database of the PBX is also presented. The type of the call (whether it is incoming our outgoing, whether local, transiting to another PBX, or connected to the public network) is also identified. Indeed, in a network of PBXs, it might be the case that not all PBXs are connected to the public network. A call can originate from a "peripheral" PBX, transit to a "central"–"gateway" PBX and then reach the public network from there. There are at least six different call types, in three categories: local (intra-PBX), incoming, and outgoing. Local calls can take place in the same PBX or towards another PBX of the network. Incoming calls might be terminating to a set that is placed in the PBX that is directly connected to the public network, or they can be towards a set placed in a "peripheral" PBX that is not directly connected to the public network. In that case the call has to transit through the "gateway" PBX. The

```
----[/DHS3dyn/account/TAXATGHP.DAT : Ticket number ████ ]----------------
<00>              TicketVersion = ED5.1
<01>               CalledNumber = █████████
<02>              ChargedNumber = █████████
<03>            ChargedUserName = █████████
<04>          ChargedCostCenter = █████
<05>             ChargedCompany =
<06>           ChargedPartyNode = 105
<07>                 Subaddress =
<08>              CallingNumber =
<09>                   CallType = PublicNetworkIncomingCallToPrivateNetwork
<10>                   CostType = ISDNCircuitSwitchedCall
<11>                EndDateTime = 20080826 01:35:40
<12>                ChargeUnits = 0
<13>                   CostInfo = 0
<14>                   Duration = 417
<15>              TrunkIdentity = 360
<16>         TrunkGroupIdentity = 1
<17>                  TrunkNode = 101
<18>          PersonalOrBusiness = Normal
<19>                 AccessCode =
<20>         SpecificChargeInfo =
<21>            BearerCapability = Speech
<22>              HighLevelComp = Unspecified
<23>                 DataVolume = 0
<24>             UserToUserVolume = 0
<25>             ExternFacilities =
<26>             InternFacilities = BasicCall
<27>              CallReference = 0
<28>               SegmentsRate1 = 0
<29>               SegmentsRate2 = 0
<30>               SegmentsRate3 = 0
<31>                    ComType = Voice
<32>          X25IncomingFlowRate = Unspecified
<33>          X25OutgoingFlowRate = Unspecified
<34>                    Carrier = 0
<35>          InitialDialledNumber = ████
<36>             WaitingDuration = 1
<37>         EffectiveCallDuration = 417
<38>        RedirectedCallIndicator = 0
<39>                StartDateTime = 20080826 01:28:43
<40>          ActingExtensionNumber =
<41>            CalledNumberNode = 9999
<42>           CallingNumberNode = 9999
<43>      InitialDialledNumberNode = 9999
<44>       ActingExtensionNumberNode = 9999
<45>       TransitTrunkGroupIdentity = 32767
<46>               NodeTimeOffset = 0
```

Fig. 5.8   A typical PBX call detail record

same distinction is made to outgoing calls. There are calls from the "gateway" PBX that directly get to the public network and outgoing call from a "peripheral" PBX that have to transit through the "gateway" PBX. The situation can get even more complicated with PBX nodes in different time zones, where the respective time difference can also be included in the log. Regarding trunk lines, there might be many different ones, each one holding different circuits and capacities (e.g., an ISDN-E1 trunk has 30 circuits). As such the trunk group along with the exact circuit used is logged. Any network features used (such as caller ID restriction) are shown too. In case of an incoming call, if the initial dialed number was forwarded, then the call gets registered to the final destination, but there is information about the number that was initially dialed too.

The charged units and cost of the call can be included in the log, as long as the telecom provider–carrier includes that information in the signaling exchanged. It is also possible to use cost-centers to help better understand the telecommunication expenses of the organization. Phones are grouped to cost-centers (one for each department), each center holding a number of phones. This way, the administrator can easily calculate the overall cost of calls of each department. One problem for the forensics analyst here is that most administrators choose to log only calls towards or from the public network. Intra-PBX calls are not logged since their volume is much higher while they are free. At the same time, there are unlimited local calls (intra-city) dial plans offered from many telecom operators. It is again possible that such calls are not logged.

From a forensics point of view, the examiner has to find out who originated that call. If the call was an internal one, then there is immediate information pointing to the internal user. As discussed earlier, a public phone such as one found in an elevator or a meeting room might well be the "culprit." However, that internal phone might have been used in a call forwarding scenario. In that case, the internal phone is call forwarded to an external destination and the fraudster is calling it (with a local or free, in case of 800 call) to get connected to his next destination. Depending on the call logging system, there might be an indication that the registered call is part of a transferred call. The examiner has to search the call logs for incoming calls that started at the same time and have the same duration. In most cases, the record of the call that was forwarded will have an indication about the call being redirected. Let us give an example to make it more clear. Fraudster is calling at 22:05:10 from payphone with number A to a PBX extension B that is forwarded (with an immediate forward, no ring at all) to an international destination number C. Number C rings for 6 s and then it is answered, with the call holding till 22:10:11. The suspicious record that immediately strikes is the record pointing to an outgoing call from set B to number C. This record has a waiting duration of 6 s and effective call duration of 4 min and 55 s (22:05:10 to 22:10:11 is 5 min and 1 s; subtracting 6 s of ringing time before the call being answered accounts for 4 min and 55 s of net calling time). The analysis must not stop here however. She must examine calls that have the same (more or less) start date and time. Doing so, she finds that there is another incoming call, from 22:05:10 to 22:10:11, that was immediately answered with a whole duration of 5 min and 1 s). The originating number points to an external number, while the internal number is the one making the international call. Luckily, many systems, mark the respective records with a flag, implying that the call was initially forwarded from another number. The situation can get more complex if set B does not have an immediate call forward, but a forward on no answer. In that case, the examiner has to search for incoming calls that were placed earlier (so much earlier as the no answer time duration set before a transfer is initiated). The overall duration of that call would be longer but the effective duration would be equal to the duration of the call to the international destination. It is immediately apparent that the process is easily handled with specific software performing the mining in the records.

5.9   Real-Time Data

5.9.1   In General

As was earlier stated, call records get created only after the call ends. For the duration of the call, no record is present. This makes it impossible to trace with call records an incident that takes place at real time. Luckily, there are commands and tools that can help in that.

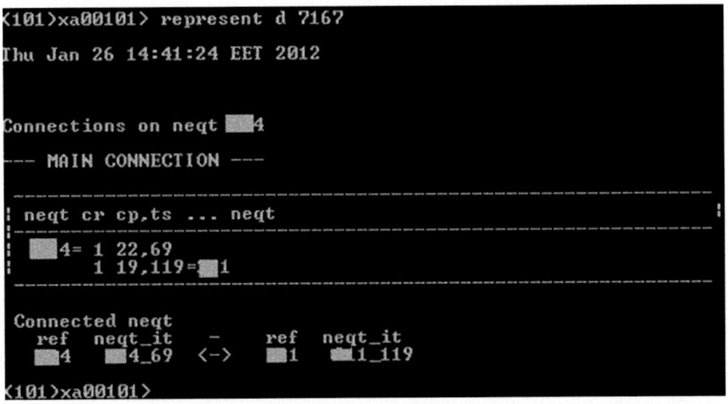

```
<101>xa00101> represent d 7167
Thu Jan 26 14:41:24 EET 2012


Connections on neqt ▆4
--- MAIN CONNECTION ---

----------------------------------------------------------------------------
: neqt cr cp,ts ... neqt                                                    !
:---------------------------------------------------------------------------
: ▆4= 1 22,69
:        1 19,119=▆1
----------------------------------------------------------------------------

Connected neqt
    ref   neqt_it    -     ref   neqt_it
    ▆4    ▆4_69   <->    ▆1   ▆1_119
<101>xa00101>
```

Fig. 5.9 Trace of connected equipment

5.9.2   Equipment Connection

Figure 5.9 shows a command that describes the path of the connection of two extensions during a call. We can see that the extension with an ID that ends in "4" is
located in board 1–22 and using timeslot 69 is further connected to the extension
that ends in 1, that is located in board 1–19 using timeslot 119. Since these two
boards are in the same rack, they are connected using the backbone of the PBX rack
without any further board in the way. It is interesting to note that although we are
discussing about set number 7167, internally this is mapped to an "equipment"
number (the one that is masked and ends in "4"). This is a way some PBXs use to
map all of their parts (extensions, boards, circuits, etc.) with a single ID.

5.9.3   Trunk Lines Data

For a call that spans outside the current PBX trunk lines are needed. Trunk lines are
the means that connect the PBX with other PBXs and the public network. Their
status can be monitored with commands that output something like what is seen in
Fig. 5.10. There we have two PRI Euro ISDN lines, totaling 60 channels. As we can
see there are some channels busy (B) and most of them free (F) at that moment.

Getting into more details, Fig. 5.11 depicts debug data for every single channel
of a trunk, along with maintenance info. Among other things there is the status of
the trunk (operating or not), information about the total number of channels occu-
pied, and whether they are incoming or outgoing. Each channel has its own number
that is mapped to an ID, while there is also the "name" of trunk appointed by the
administrator of the system as well as different administrative data.

```
+=================================================================================+
: T R U N K      S T A T E        Trunk group number : 1                          :
:                                 Trunk group name    : ▓▓▓                        :
:                                 Number of Trunks    : 60                         :
+---------------------------------------------------------------------------------+
: Index :    1    2    3    4    5    6    7    8    9   10   11   12   13          :
: State :    B    B    B    B    B    B    B    B    B    B    B    B    B          :
+---------------------------------------------------------------------------------+
: Index :   14   15   16   17   18   19   20   21   22   23   24   25   26         :
: State :    F    F    F    F    F    F    F    F    F    F    F    F    F          :
+---------------------------------------------------------------------------------+
: Index :   27   28   29   30   31   32   33   34   35   36   37   38   39         :
: State :    F    F    F    F    B    B    B    B    B    F    F    B    B          :
+---------------------------------------------------------------------------------+
: Index :   40   41   42   43   44   45   46   47   48   49   50   51   52         :
: State :    F    F    F    F    F    F    B    F    F    F    F    F    F          :
+---------------------------------------------------------------------------------+
: Index :   53   54   55   56   57   58   59   60                                  :
: State :    F    F    F    F    F    F    F    F                                   :
+---------------------------------------------------------------------------------+
: F:   Free       :   B:  Busy      :  Ct: busy Comp trunk   :  Cl: busy Comp link :
: WB:  Busy Without B Channel:  Cr: busy Comp trunk for RLIO inter-ACT link        :
: WBD: Data Transparency without chan.:  WBM: Modem transparency without chan.     :
: D:   Data Transparency       :   M: Modem transparency                          :
+=================================================================================+
```

Fig. 5.10   Trunk lines status (showing a 2 PRI Euro ISDN trunk)

5.9.4   Signaling Data

The actual flow of communication in the trunk lines is controlled with the relevant signaling. One of the most important and powerful monitoring tools, therefore, is the tool to monitor the signaling between the PBX and the other PBXs or the central office. Figure 5.12 shows the exchange of ISDN messages, where a given phone calls another phone, in a given channel and then hang-ups. We can find information both for the PBX itself (e.g., board numbers and circuit numbers) and for ISDN signaling information that is exchanged with the central office. Each trace has a time stamp and a serial number. The volume of signaling numbers is far greater that the number of call records, since for each call the whole interexchange of messages is logged. This makes it practically impossible to log this signaling for a long time. Rather, it can be used in real time to monitor specific lines, either tracing the flow of calls before the call records are created or in order to troubleshoot a given problem.

One very important feature in both PBXs and home lines is the malicious call feature where the user who has received a malicious call can "stamp" the call as such and notify the central office that the call should specifically be logged in order to preserve the evidence to allow him to take further legal steps. This is usually a paid service. Speaking of ISDN signaling traces, as Fig. 5.13 shows, if the PBX owner has not paid the operator for the service then the respective message (not subscribed) appears in the traces log.

5.10   Extensions' Data

Besides trunk lines, many forensics data stem from an action performed by an extension. PBX extensions (namely phones) have evolved from old analog phones to expensive digital electronic devices with complex electronic circuits. It is therefore

```
****************** data in Trunk_Group structure ******************

********          data FAISCEAU_NON_RESIDENT
nomfais           =
discrLogId        = 0        ton_a_prise  = 1           em_repfix  = 0
lignpriv          = 0        reservop     = 0           reserauto  = 0
faisrecherche     = 0        ftranscom    = 0           frondier   = 0
typjo :      (6) => T2
fais_suiv         = -1       nbchifem     = 0           tab_proto  = -1
var ISDN = 2
dupplicated IG ; <= node_number     = 255
network_number = 15
faisc_reg_sig = 0
special_it_par_quantum = 1
cat_robinet_in  = 10, cat_robinet_out = 10
Priority ===> Level= 0, Mode= 0, Preemption= 0
mpt1343 = 0, rasfo = 1 rerouting = 0
********          data ARTERE(link)
cat_signa = 19   ch_canalb = 0    deborde_it = 1   access_turn = 1 mode_reseau = 0
+------------------------------------------------------------------------------
|                           NEW ocupjonc
+------------------------------------------------------------------------------
| i = 0, min = 0, max = 29
| (num_crist - num_cpl - num_term) = (0-25-0)
| dernier it pris = 0,  monlap = 40,  mode_reseau = 0  nbr_jonc_cree = 30
| nombre_jonc_occ :  depart = 7  arrivee = 0  mixte = 7
| it_reservees     :  depart = 0  arrivee = 0  mixte = 30
| it_max_Q0        :  depart = 0  arrivee = 0  mixte = 30
| it_max_Q1        :  depart = 0  arrivee = 0  mixte = 0
| acces_niveau2 = CONNECT2
+------------------------------------------------------------------------------
| hs ! res ! occ !nulog!trans!neqtdyn!E64 RN64 EN64! OUPN !  neqph  ! adr
+------------------------------------------------------------------------------
! LIB ! non ! DEP !  0 !  1 !  22 ! 1   0   0 !  0  !  1322 !0-  0-1
! LIB ! non ! lib !  1 !  1 !   1 ! 0   0   0 !  0  !  1323 !0-  -2
! LIB ! non ! DEP !  2 !  1 !  24 ! 1   0   0 !  0  !  1324 !0-  0-3
! LIB ! non ! DEP !  3 !  1 !  25 ! 1   0   0 !  0  !  1325 !0-  0-4
! LIB ! non ! lib !  4 !  1 !   1 ! 0   0   0 !  0  !  1326 !0-  -5
! LIB ! non ! lib !  5 !  1 !   1 ! 0   0   0 !  0  !  1327 !0-  -6
! LIB ! non ! DEP !  6 !  1 !  28 ! 1   0   0 !  0  !  1328 !0-  0-7
! LIB ! non ! lib !  7 !  1 !   1 ! 0   0   0 !  0  !  1329 !0-  -8
! LIB ! non ! DEP !  8 !  1 !  30 ! 1   0   0 !  0  !  1330 !0-  0-9
! LIB ! non ! lib !  9 !  1 !   1 ! 0   0   0 !  0  !  1331 !0-  -10
! LIB ! non ! lib !  0 !  1 !   1 ! 0   0   0 !  0  !  1332 !0-  -11
! LIB ! non ! DEP !  1 !  1 !  33 ! 1   0   0 !  0  !  1333 !0-  0-12
! LIB ! non ! DEP !  2 !  1 !  34 ! 1   0   0 !  0  !  1334 !0-  0-13
! LIB ! non ! lib !  3 !  1 !  -1 ! 0   0   0 !  0  !  1335 !0-  -14
! LIB ! non ! lib !  4 !  1 !  -1 ! 0   0   0 !  0  !  1336 !0-  -15
! LIB ! non ! lib !  5 !  1 !  -1 ! 0   0   0 !  0  !  1337 !0-  -17
! LIB ! non ! lib !  6 !  1 !  -1 ! 0   0   0 !  0  !  1338 !0-  -18
! LIB ! non ! lib !  7 !  1 !  -1 ! 0   0   0 !  0  !  1339 !0-  -19
! LIB ! non ! lib !  8 !  1 !  -1 ! 0   0   0 !  0  !  1340 !0-  -20
! LIB ! non ! lib !  9 !  1 !  -1 ! 0   0   0 !  0  !  1341 !0-  -21
! LIB ! non ! lib !  0 !  1 !  -1 ! 0   0   0 !  0  !  1342 !0-  -22
! LIB ! non ! lib !  1 !  1 !  -1 ! 0   0   0 !  0  !  1343 !0-  -23
! LIB ! non ! lib !  2 !  1 !  -1 ! 0   0   0 !  0  !  1344 !0-  -24
! LIB ! non ! lib !  3 !  1 !  -1 ! 0   0   0 !  0  !  1345 !0-  -25
! LIB ! non ! lib !  4 !  1 !  -1 ! 0   0   0 !  0  !  1346 !0-  -26
! LIB ! non ! lib !  5 !  1 !  -1 ! 0   0   0 !  0  !  1347 !0-  -27
! LIB ! non ! lib !  6 !  1 !  -1 ! 0   0   0 !  0  !  1348 !0-  -28
! LIB ! non ! lib !  7 !  1 !  -1 ! 0   0   0 !  0  !  1349 !0-  -29
! LIB ! non ! lib !  8 !  1 !  -1 ! 0   0   0 !  0  !  1350 !0-  -30
! LIB ! non ! lib !  9 !  1 !  -1 ! 0   0   0 !  0  !  1351 !0-  -31
+------------------------------------------------------------------------------
```

Fig. 5.11 More info on a given trunk line

possible that data are stored not only in the main HDD of the PBX's CPU board, but also in the memory chips of these extensions. Apart from classical PBX phone sets, VoIP hardphones store considerable information in their own memory and therefore might be advantageous to perform a forensics analysis there too. Two memory types are used, the NAND-flash and the NOR-flash. The actual data stored depend on the manufacturer and the architecture of the system. Beyond the current data stored, older, or even deleted data can be found in the "depths" of the phone's memory. Such information is conditionally possible to be (partially or totally) recovered after

```
(772794:000018) 1444: Send_IO1 (link-nbr=17, sapi=0, tei=0) :
long: 41  desti: 0  source: 15  cryst: 1  cpl: 24  us: 8  term: 0  type a5
tei: 0  <<<<  message sent : SETUP [05]     Call ref :    39 06


IE:[04] BEARER_CAPABILITY (l=3) 80 90 a3
IE:[6c] CALLING_NUMBER (l=12)  -> 01  81  Num : 2▮▮▮▮▮▮7
IE:[7d] HLC (l=2) 91 81



(772796:000019) Physical-Event :
long: 27  desti: 0  source: 0  cryst: 1  cpl: 24  us: 0  term: 0  type a5
tei: 0  >>>>  message received : SETUP ACK [0d]  Call ref :   b9 06


IE:[18] CHANNEL (l=3)  a9 83 89 -> T2 : B channel 9 exclusive
IE:[1e] PROGRESS_ID (l=2) 82 88



(772796:000020) 1444: Send_IO1 (link-nbr=17, sapi=0, tei=0) :
long: 32  desti: 0  source: 15  cryst: 1  cpl: 24  us: 8  term: 0  type a5
tei: 0  <<<<  message sent : INFORMATION [7b]  Call ref :   39 06


IE:[70] CALLED_NUMBER (l=11)  -> 81  Num : ▮▮▮▮▮▮52
[a1] Sending complete



(772797:000021) Physical-Event :
long: 18  desti: 0  source: 0  cryst: 1  cpl: 24  us: 0  term: 0  type a5
tei: 0  >>>>  message received : CALL PROC (02)  Call ref :   b9 06


(772828:000022) 1444: Send_IO1 (link-nbr=17, sapi=0, tei=0) :
long: 26  desti: 0  source: 15  cryst: 1  cpl: 24  us: 8  term: 0  type a5
tei: 0  <<<<  message sent : DISCONNECT [45]  Call ref :   39 06

IE:[08] CAUSE (l=2) 81 90 -> [90] NORMAL CALL CLEARING
IE:[1e] PROGRESS_ID (l=2) 81 88
```

Fig. 5.12 ISDN traces, showing calling and called number

```
(684537:000193) 1444: Send_IO1 (link-nbr=17, sapi=0, tei=0) :
long: 29  desti: 0  source: 15  cryst: 1  cpl: 24  us: 8  term: 0  type a5
tei: 0  <<<<  message sent : FACILITY [62]  Call ref :    80 bd


IE:[1c] FACILITY (l=9)
  [91] Discriminator of supplementary service applications
  [a1] INVOKE (l=6):
      Invoke Ident. : 0003 (3)
      OP: 0003 (3)



(684538:000194) Physical-Event :
long: 29  desti: 0  source: 0  cryst: 1  cpl: 24  us: 0  term: 0  type a5
tei: 0  >>>>  message received : FACILITY [62]  Call ref :   00 bd


IE:[1c] FACILITY (l=9)
  [91] Discriminator of supplementary service applications
  [a3] ERROR (l=6):
      Invoke Ident. : 0003 (3)
      [02] General error (l=1) NotSubscribed (0)
```

Fig. 5.13 ISDN signaling trace showing a non-subscribed service being asked for

deletion. The examiner is called to extract these data without altering them [4]. For this purpose he can resort to the memory dump byte to byte for NOR memory (or page to page for NAND type memory). This is achieved using special software and hardware tools that "clone" the memory contents. This way she can receive a complete data image in physical layer. The whole process is particularly complicated from technical point of view because the data are unstructured, and they have to be translated in a specific file system.

There is no standardized way or a common contact for the connection (each manufacturer uses a different one even between his own models) while it is often required to access special internal contacts, found in the printed circuit board of the phone. One of these contacts is Joint Test Action Group (JTAG) described in IEEE 1149.1 standard [5]. The functionality offered is a very powerful feature. It is a special debug interface initially used to test the printed circuit boards of electronic devices. It further evolved to allow checking, monitoring, and debugging of embedded systems. For forensics use, it can undertake the memory dump but it is difficult to find the specific test points, and usually documentation is proprietary information not readily available. Despite the advantages of the memory dump method, certain problems manifest. The basic problem is that there is no way to detect if external changes have taken place in the flash memory (therefore the data may have been modified).

If no other method can be used (e.g., because the phone is partially or totally destroyed), it is also possible to detach–desolder the integrated memory circuits using special precision surface mount device (SMD) soldering/desoldering stations. Following that, external memory dump takes place using the right hardware tools. This process guarantees that no data "infection" happens since the phone remains unconnected from the PBX. It faces, however, the serious danger of the complete circuit destruction during that delicate process of detachment/desoldering. Moreover, the extension should be disassembled in order to extract the integrated memory circuit. Considering these difficulties, this method is the least preferred and the one that the analyst will ultimately resort to if nothing else seems to be working.

In a more convenient way, most of the evidence needed for a case is found in the HDD of the PBX's CPU and can be acquired using the right commands. One such command lists all the configuration details and operation details of a given set. Figure 5.14 shows a partial output (out of many pages of information) for a given set (number 7167). The reader can note that the specific PBX uses a UNIX–Linux like operating system where commands such as "grep" can be used. Information is masked in the screenshot, but contains the name of the user, the last dialed number in the redial memory (06945-masked), the last called number in the missed calls memory (06945-masked), whether mail notification is switched on or off, the volume level of the ringer, the volume level of the earpiece, the distinctive ringing pattern chosen by the user, and so on.

Specifically for sets with programmable or static keys, there is the option to retrieve their contents. From a forensics point of view, numbers stored in the memories of the phone can link a suspect to another one, even if he didn't ever call that number from the given telephone. It might also be possible that other codes such as PINs are stored there, which can be used in the specific case. Figure 5.15 shows a listing of all the keys (28) of a given set. Keys with programmed numbers have again been masked.

```
<101>xa00101> ███████ d 7167  | grep num
neqt=███;                    numan [1..8] = 7167      nomannu[length=15] = ████████
numass = ████?               numadec = 0             nulog=███2
mask_number       = 0
text_msg_number = 8
numeenr =[1..15] =      10   6   9   4   5  ████████████████  0   0   0   0
nbchmemo = 0    numdest =7167    renvoi = 4    typdest = 4
pt_rall->numdest_sec =7167        pt_rall->renvoi_sec = 4 pt_rall->typdest_sec = 4
pt_rall->nivaudio_comb_num = 5   pt_rall->nivaudio_comb_anal = 2
pt_rall->idx_data = 0   pt_rall->lien_ass = -1   pt_rall->num_son  = 7
pt_rall->numan_secdeb =
numenrco =[1..15] =     10   6   9   4   5  ████████████████  0   0   0   0
Mailstatus.Textnumber.Number    = 0
Mailstatus.Textnumber.Age       = 6
Mailstatus.Cabacknumber.Number  = 0
Mailstatus.Cabacknumber.Age     = 6
numerovestal [1..15] =     0   0   0   0   0   0   0   0   0   0   0   0   0   0
  0
numtougarde = 0 neqtparc = -1    num_secretaire = ████?
<101>xa00101> _
```

Fig. 5.14 A digital set's memory dump

```
Key  2:   Multi MCDU          --->       Key content: 4801
               Dom   0  CH DATA : Tsl_CH_Local_Station ID = ██

keytype: 255
genrtou: 11    inf_num: -1  lp_occup: 0  etat_led: 45   ligne_ACD: 0
cor_ra: 0    cor_gar: 0    cor_rapgar: 0    cor_pbx: 0    cor_reser: 0
cor_acd: 0    cor_pre: 0    cor_gar_rec: 0    sonsurson: 0   distri_call: 0
autoriz_gar_rec: 0    prio_appel: 0    pri_cour_line: Act=0-Level=0
appel_tanden: 0    appel_presente: 0    cor_pbx_network: 0    secondary_inter: 0
order_MLA: 0       no_tou_MLA_Prim: 0
sonDiffere: 0      FinImpSonDif: 0      FromHoldButton: 0      OnlyPrivateHold: 0
tou_supenquiry_ACD: 0


Key  3:   SPK key             Camp_On_Control
Key  4:   Forward on ringing --->       Key content: ███8
               Dom   0  CH DATA : Tsl_CH_Local_Station ID = ███

Key  5:   Headset
Key  6:   Prog. Key           Empty key
Key  7:   Prog. Key           Empty key
Key  8:   SPK key             REPERT
Key  9:   Prog. Key           Empty key
Key 10:   SPK key             BOOKING
Key 11:   ISDN Key            --->       Key content: ████
               Dom   0  CH DATA : Tsl_CH_Local_Station ID = ███

Key 12:   Prog. Key           Empty key
Key 13:   Prog. Key           Empty key
Key 14:   Prog. Key           Empty key
Key 15:   Prog. Key           Empty key
Key 16:   ISDN Key            --->       Key content: 08071122
               Dom   0  Trunk Group : Tsl_Business_TG_Overlap_Pfx ID = 1 , ADT #
0, PrivRoute 0

Key 17:   Prog. Key           Empty key
Key 18:   Prog. Key           Empty key
Key 19:   SPK key             REDIAL_MEM
Key 20:   Prog. Key           Empty key
Key 21:   SPK key             MAIL key
Key 22:   SPK key             ISDN
Key 23:   SPK key             REDIAL_BIS
Key 24:   SPK key             Loudspeaker key
Key 25:   SPK key             Loudspeaker adjustement -
Key 26:   SPK key             Loudspeaker adjustement +
Key 27:   SPK key             Mute & Intercom key
Key 28:   SPK key             Handsfree key
```

Fig. 5.15 Programmable keys contents of a given set

Getting the numbering plan of the PBX, the examiner is in position to know what service exactly did the user ask for. Indeed, the numbering plan describes all possible key sequences in the PBX and their meaning. Figure 5.16 depicts a partial list of features and their respective enabling codes.

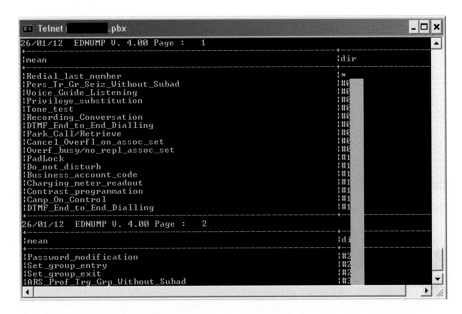

Fig. 5.16 Numbering plan and features

5.11   Evidence Stored Outside the PBX

As we have seen, a management station-console could be connected (directly or via
network) to the PBX. Consequently, not only the PBX but also the hard disk and the
computer memory might contain data from the interaction of the computer with the
PBX. For low end PBXs, the whole management suite might be installed in the
management PC, holding also backups and other valuable evidence. This means
that the examiner's work is not limited to the PBX but it also continues to the computer. We won't extend further our analysis because digital evidence extraction
from computers is a whole subject by itself.

Apart from data that the expert can extract from the PBX and the management
PC, the network operator maintains logs of calls with the respective CDRs. This
information can be correlated to the call logs evidence extracted from the PBX
itself. At this point, different retention laws mandate the duration that the provider–
carrier has to keep these data.

Speaking of telecom operators, it might be possible that the PBX uses not only a
single one for its calls but more, either for redundancy or for economy reasons. In
that case, users might select the desired carrier by pressing a specific code before the
call. Or, the process can be automated with automated routing selection (ARS) features where the PBX decides what is the best (usually the most economic at that
specific time) carrier to select for a given call. This, again, complicates the situation
since the examiner has to find CDRs from all carriers involved. In addition to that,
fraudsters, using carrier selection dialing codes, might chose a third carrier to place

the call. That could sometimes blur the CDRs from the main carriers the PBX is served from. Indeed, in such an occurrence, it is possible that the CDR of the original carrier does not contain the final destination reached, but rather an indication that the call was served by another carrier, along with the respective cost.

5.12 Conclusion

Given the increased criminal activity targeting PBXs, forensics is an essential asset for law enforcement and an important topic to discuss. Digital forensics analysis is a permanent and difficult fight against criminals. Their modus operandi was presented in this chapter for the user to be able to understand the forensics procedures involved when analyzing an incident. At the same time, dozens of different PBX models exist, with different characteristics and interfaces. Commands and tools that can help in the evidence extraction can be undocumented or difficult to use. Moreover, a relative lack of forensics standards and exceptional complexity in the providers' networks make things worse, while the examiner is often called to work under pressure and narrow time limits.

Beyond the technical details and complexity, nontechnical issues play an equally important role. The digital forensics examiner should be constantly up to date with the field's technological developments and follow specific and established procedures. Once again, education is what will help combine the technical with the procedural part so that the work of the analyst can substantially help the proper trial of the case.

This is why, in this chapter, we attempted to describe the methodology and procedures encountered in collecting, extracting, analyzing, and presenting evidence from PBXs, from both the theoretical and the technical side. Indeed, in case the user of a phone is involved in criminal/illegal activities, data from the PBX can prove evidence and can play an important role for the law enforcement and judicial authorities. The other way around, if a PBX is victim of economic crime, the forensics procedure can help find the fraudsters.

References

1. Harrill DC, Mislan RP (2007) A small scale digital device forensics ontology. Small Scale Digital Device Forensics J 1(1)
2. International Organization on Computer Evidence guidelines for best practice in the forensic examination of digital technology. http://www.ioce.org/core.php?ID=5
3. ACPO Good Practice Guide for Computer-Based Evidence. ACPO. http://www.7safe.com/electronic_evidence/ACPO_guidelines_computer_evidence_v4_web.pdf
4. Breeuwsma M, de Jongh M, Klaver C, van der Knijff R, Roeloffs M (2007) Forensic data recovery from flash memory. Small Scale Digital Device Forensics J 1(1)
5. IEEE (2001) IEEE 1149.1, IEEE standard test access port and boundary scan architecture

Chapter 6
Conclusions

In this book we tried to raise users' and administrators' awareness in regard to security and privacy threats targeting PBXs. Focused on practical issues and skipping theoretical analysis of algorithms and standards we presented the vulnerabilities and the way malicious users take advantage of them.

Modern societies are highly dependent on communication and loss of availability issues can cause a great deal of inconvenience or even life-threatening situations. Furthermore, important personal and business privacy intrusions include not only the interception of voice but also data such as faxes. Industrial espionage is a typical example. In another front, telecom fraud, via PBXs accounts for billion of losses per year, exploited not only by malicious hackers but also by organized crime and terrorists.

Given their importance, PBXs are part of a nation's critical infrastructure and they should be adequately protected. Securing them is a process that involves both technical and nontechnical steps. We therefore tried to cover as many steps as possible offering relevant advice and know-how in the respective chapter.

Even for protected PBXs, attackers will eventually find new ways of abusing the systems, while the research community will continuously actively challenge their security. As such, for cases where an incident has already taken place, the forensics chapter provided the insight to the necessary methodology to investigate it as well as the tools that should be used to extract evidence.

As a closing remark, the technology itself, either PBX or VoIP, is practically of little importance. Each system has its own advantages and disadvantages in regard to security. Both systems, however, face the same threats. Hackers and malware with a similar modus operandi could effectively target any platform. This is why, in this book, we focused mostly on problems that are common for most systems and on universal solutions that can equally be applied. Along with a proposal for a centralized and focused project to secure PBXs, the book will hopefully form the basis for an actual security improvement, implemented by the various stakeholders.

© Springer International Publishing Switzerland 2016
I.I. Androulidakis, *VoIP and PBX Security and Forensics*, SpringerBriefs
in Electrical and Computer Engineering, DOI 10.1007/978-3-319-29721-7_6

About the Author

Dr. Iosif I. Androulidakis has an active presence in the ICT security field having authored more than 90 publications (including six books) and having presented more than 120 talks and lectures in international conferences and seminars in 20 countries.

Holding two Ph.D.s his research interests focus on security issues in PBXs (private telephony exchanges), where he has 20 years of experience, as well as in mobile phones and embedded systems. Part of his research has led to the granting of five patents. During his career, he collaborated with telecom operators, national police Cybercrime departments, the European Police Academy (CEPOL), the European Public Law Center, the Southeastern Europe Telecommunications & Informatics Research Institute, Universities and Research centers, Vocational training institutes, and the Media and Private security consulting firms.

Dr. Androulidakis has also acted as a reviewer in an extended array of scientific conferences and journals, as a Programme Committee member in 26 conferences and as a chairman in eight conference sessions. Finally, he is a certified ISO9001 (Quality Management System) and ISO27001 (Information Security Management System) auditor and consultant.

© Springer International Publishing Switzerland 2016

I.I. Androulidakis, *VoIP and PBX Security and Forensics*, SpringerBriefs in Electrical and Computer Engineering, DOI 10.1007/978-3-319-29721-7